Student Solutions Manual
for Blei and Odian's
General, Organic, and Biochemistry
Second Edition

Mark Dadmun
The University of Tennessee

W. H. Freeman and Company
New York

© 2006 by W. H. Freeman and Company

All rights reserved.

Printed in the United States of America

ISBN: 0-7167-7373-2
EAN: 9780716773733

Second printing

W. H. Freeman and Company
41 Madison Avenue
New York, NY 10010
Houndmills, Basingstoke
RG21 6XS England

Table of Contents

Preface

Chemistry plays an important role in many aspects of everyday life, from cooking to personal care products to health care. Understanding how chemical transformations occur provides you with knowledge that can be called upon and utilized in the future to make smart consumer decisions, to fix problems, and to perform important job-related tasks. This last aspect is particularly true if you are planning a profession in the health sciences—a field in which an understanding of how medications work and how the body functions, *from a chemical standpoint*, is essential for a successful and prosperous career.

General, Organic, and Biochemistry by Blei and Odian is designed to provide an introduction to the concepts of chemistry for students who are planning a career in the health sciences. The book is separated into three sections corresponding to the title: Chapters 1–10 introduce the important concepts of general chemistry, Chapters 11–17 provide the fundamentals of organic chemistry, and Chapters 18–26 explain the chemistry of the body—biochemistry.

This manual provides the complete solutions to the odd-cumbered exercises at the end of each chapter of *General, Organic, and Biochemistry*. The explanations were written to show you the thought process required to solve the problem. Additionally, the explanations exemplify the mathematical level and problem-solving skills that are required for mastery of the material in that chapter.

Although the answers to these questions can be found in the back of the text, the purpose of this manual is to explain *how* the answers are derived. It is important to use this manual as a check of your work. Though it may be tempting to peruse the manual before working the problems, you will only learn the material (which is the goal, isn't it?) if you work the problem first, and then refer to the manual.

This manual also can be used for *Organic and Biochemistry*, by Blei and Odian. This sister text synopsizes the concepts of general chemistry into two chapters and then proceeds to cover the important concepts of organic chemistry and biochemistry. It is considered a corresponding text to *General, Organic, and Biochemistry* because Chapters 1 and 2 summarize material from Chapters 1–10 of *General, Organic, and Biochemistry*, and Chapters 3–18 correspond directly to Chapters 11–26 of *General, Organic, and Biochemistry*. Thus, Chapters 11–26 of this manual also provide the worked solutions to the odd-numbered end-of-chapter exercises of Chapters 3–18 of *Organic and Biochemistry*. The solutions to all end-of-chapter exercises in Chapters 1 and 2 of *Organic and Biochemistry* are provided in an Appendix at the end of this manual.

This manual also can be used for *An Introduction to General Chemistry*, by Blei and Odian, because Chapters 1–10 of *An Introduction to General Chemistry* correspond directly to Chapters 1–10 of *General, Organic, and Biochemistry*. Thus, Chapters 1–10 of this manual provide the worked solutions to the odd-numbered end-of-chapter exercises of *An Introduction to General Chemistry*.

Finally, I am indebted to a number of people who helped in the preparation of this manual. Most importantly, however, I would like to thank my family my wife Jayne and my four children, Ryan, Catherine, Maggie, and Ben for their love, patience, and support during the writing of this manual.

Chapter 1

The Language of Chemistry

1.1   A *physical* property is one that a substance can undergo without changing its identity. A *chemical* property is one that describes how a substance can react with another substance to form a new substance.

(a) Oxygen loses its identity; therefore, this is a chemical property.

(b) When oxygen boils, it transforms from a liquid into a gas but does not change identity; therefore, this is a physical property.

(c) When oxygen dissolves in water, it mixes with water on a molecular level but does not change its identity; therefore, this is a physical property.

(d) When oxygen reacts with hydrogen to form water, oxygen loses its identity; therefore, this is a chemical property.

1.3   (a) Air is a mixture of oxygen and nitrogen.

(b) Mercury is a pure substance found on the periodic table.

(c) Aluminum foil is made only of aluminum; therefore it is a pure substance.

(d) Table salt is NaCl, a pure compound that cannot be separated.

1.5   Length is measured in meters; mass in kilograms; temperature in Kelvins; time in seconds; amount of substance in moles.

1.7   (a) $630 \text{ m} \times \left(\dfrac{1 \text{ km}}{1000 \text{ m}} \right) = 0.630 \text{ km}$

(b) $1440 \text{ ms} \times \left(\dfrac{1 \text{ s}}{1000 \text{ ms}} \right) = 1.440 \text{ s}$

(c) $0.000065 \text{ kg} \times \left(\dfrac{1000 \text{ g}}{1 \text{ kg}} \right) \times \left(\dfrac{1000 \text{ mg}}{1 \text{ g}} \right) = 65 \text{ mg}$

(d) $1300 \text{ } \mu\text{g} \times \left(\dfrac{1.000 \text{ mg}}{1000 \text{ } \mu\text{g}} \right) = 1.300 \text{ mg}$

1.9   (a) $0.0945 = 9.45 \times 10^{-2}$; therefore there are three significant figures.

(b) $83.22 = 8.322 \times 10^{1}$; therefore there are four significant figures.

(c) $106 = 1.06 \times 10^{2}$; therefore there are three significant figures.

(d) $0.000130 = 1.30 \times 10^{-4}$; therefore there are three significant figures.

1.11   (a) 25.9000 g; because last three zeros are shown, all digits are significant; therefore there are six significant figures.

(b) 102 cm $= 1.02 \times 10^{2}$; therefore there are three significant figures.

(c) 0.002 m $= 2 \times 10^{-3}$; therefore there is one significant figure.

(d) 2001 kg $= 2.001 \times 10^{3}$; therefore there are four significant figures.

(e) 0.0605 s $= 6.05 \times 10^{-2}$; therefore there are three significant figures.

1.13   (a) 0.00839; move the decimal place to the right three places to get 8.39; therefore $\equiv 8.39 \times 10^{-3}$.

(b) 83,264; move the decimal place to the left four places to get 8.3264; therefore $\equiv 8.3264 \times 10^{4}$.

(c) 372; move the decimal place to the left two places to get 3.72; therefore $\equiv 3.72 \times 10^{2}$.

(d) 0.0000208; move the decimal place to the right five places to get 2.08; therefore $\equiv 2.08 \times 10^{-5}$

1.15   (a) 936,800; move the decimal place to the left five places to get 9.368; therefore $\equiv 9.368 \times 10^{5}$.

(b) 1638; move the decimal place to the left three places to get 1.638; therefore $\equiv 1.638 \times 10^{3}$.

(c) 0.0000568; move the decimal place to the right five places to get 5.68; therefore $\equiv 5.68 \times 10^{-5}$.

(d) 0.00917; move the decimal place to the right three places to get 9.17; therefore $\equiv 9.17 \times 10^{-3}$

1.17   (a) In multiplication, the number of significant figures in the answer is equal to the number of significant figures in the least exact number. In this example, 3.21 has three significant figures; therefore the answer must have three significant figures $\equiv 48.4 \text{ cm}^{2}$.

(b) 1.1 has two significant figure; therefore the answer must have two significant figures ≡ 8.4.

1.19   In the addition to two numbers, the result should have only as many digits to the right of the decimal point as the starting number with the least number of digits to the right of the decimal point:

(a) 1.23 has only two digits after the decimal point; therefore 1.23 + 12.786 = 14.02.

(b) 3.961 has only three digits after the decimal point; therefore 3.961 + 24.6543 = 28.615.

1.21   Each answer will have three significant figures because 0.800 miles has three significant figures, which is the minimum.

(a) $0.800 \text{ miles} \times \left(\dfrac{1.6093 \text{ km}}{1 \text{ mile}}\right) = 1.29 \text{ km}$

(b) $0.800 \text{ miles} \times \left(\dfrac{1.6093 \text{ km}}{1 \text{ mile}}\right) \times \left(\dfrac{1000 \text{ m}}{1 \text{ km}}\right) = 1.29 \times 10^3 \text{ m}$

(c) $0.800 \text{ miles} \times \left(\dfrac{1.6093 \text{ km}}{1 \text{ mile}}\right) \times \left(\dfrac{1000 \text{ m}}{1 \text{ km}}\right) \times \left(\dfrac{100 \text{ cm}}{1 \text{ m}}\right) = 1.29 \times 10^5 \text{ m}$

(d) $0.800 \text{ miles} \times \left(\dfrac{1.6093 \text{ km}}{1 \text{ mile}}\right) \times \left(\dfrac{1000 \text{ m}}{1 \text{ km}}\right) \times \left(\dfrac{1000 \text{ mm}}{1 \text{ m}}\right) = 1.29 \times 10^6 \text{ m}$

1.23 $3.75 \text{ gal} \times \left(\dfrac{3.7853 \text{ L}}{1 \text{ gal}}\right) \times \left(\dfrac{1000 \text{ mL}}{1 \text{ L}}\right) = 1.42 \times 10^4 \text{ mL}$

1.25 $10 \text{ km} \times \left(\dfrac{1 \text{ mile}}{1.6903 \text{ km}}\right) = 6.214 \text{ miles}$

1.27   From Table 1.5: 1 oz = 28.35 g

1.29   (a) $0.60 \text{ lb} \times \left(\dfrac{453.59 \text{ g}}{1 \text{ lb}}\right) \times \left(\dfrac{1 \text{ kg}}{1000 \text{ g}}\right) = 0.27 \text{ kg}$

(b) $0.60 \text{ lb} \times \left(\dfrac{453.59 \text{ g}}{1 \text{ lb}}\right) = 2.7 \times 10^2 \text{ g}$

(c) $0.60 \text{ lb} \times \left(\dfrac{453.59 \text{ g}}{1 \text{ lb}}\right) \times \left(\dfrac{1000 \text{ mg}}{1 \text{ g}}\right) = 2.7 \times 10^5 \text{ mg}$

1.31 $1.0 \text{ L} = 1.0 \times 10^3 \text{ mL} = 1000 \text{ cm}^3$

$\sqrt[3]{1 \times 10^3} = 10 \times 10 \times 10 \text{ cm}$

Therefore, each side of the cube whose volume is 1.0 L is 10 cm.

1.33 $17.00 \text{ mm} \times \left(\dfrac{1 \text{cm}}{10 \text{ mm}} \right) = 1.700 \text{ cm}$

Volume $= L \times W \times H = 1.700 \text{ cm} \times 6.50 \text{ cm} \times 3.25 \text{ cm} = 35.9 \text{ cm}^3$

Mass = Density $\times$ Volume $= 2.2 \; \dfrac{\text{g}}{\text{cm}^3} \times 35.9 \text{ cm}^3 = 79 \text{ g}$

1.35 $°C = (°F - 32°F) \times \left(\dfrac{5°C}{9°F} \right)$

$°C = (68°F - 32°F) \times \left(\dfrac{5°C}{9°F} \right) = 20°C$

1.37 $K = °C + 273 = 27°C + 273 = 300 \text{ K}$

1.39 $C_p = \left(\dfrac{\text{joules}}{\text{g} \times °C} \right) = \left(\dfrac{22.0 \text{J}}{32 \text{ g} \times (28°C - 18°C)} \right) = 0.069 \left(\dfrac{\text{J}}{\text{g} \times °C} \right)$

1.41 $\rho = \text{density} = \left(\dfrac{\text{mass of solution}}{\text{volume of solution}} \right)$

Volume of solution = 19.84 mL

Mass of solution = mass of flask and solution − mass of flask

22.419 g = 54.381 g − 31.962 g

$\rho = \text{density} = \left(\dfrac{22.419 \text{g}}{19.84 \text{ mL}} \right) = 1.130 \text{ g/mL}$

1.43   Use the unit-conversion method.

$8.4 \times 10^3 \text{ kJ} \times \left(\dfrac{1 \text{ g}}{16.74 \text{ kJ}} \right) = 501 \text{ g}$

$$501 \text{ g} \times \left(\frac{1 \text{ lb}}{453.6 \text{ g}}\right) = 1.10 \text{ lb}$$

1.45 $\text{Density} = \left(\dfrac{\text{mass}}{\text{volume}}\right) = \left(\dfrac{41.242 - 28.463 \text{ g}}{12.3 \text{ mL}}\right) = \left(\dfrac{12.78 \text{ g}}{12.3 \text{ mL}}\right) = 1.04 \text{ g/mL}$

1.47 $\rho = \text{density} = 19.32 \text{ g/cm}^3$

$$2.416 \text{ kg} = 2416 \text{ g} \times \left(\frac{1 \text{cm}^3}{19.32 \text{ g}}\right) = 125.1 \text{ cm}^3$$

1.49   (a) Water changes into hydrogen gas and loses its identity; therefore, this is a chemical property.

(b) Lead does not change its identity; therefore, this is a physical property.

(c) These properties of lead do not change its identity; therefore, this is a physical property.

(d) Lead changes its identity and properties; therefore, this is a chemical property

1.51   (a) $9{,}620{,}000 \text{ kg} = 9.62 \times 10^6 \text{ kg}$

(b) $54{,}870 \text{ days} = 5.487 \times 10^4 \text{ days}$

(c) $253 \text{ milliseconds} \times \left(\dfrac{1 \text{s}}{1000 \text{ ms}}\right) = 0.253 \text{ s} = 2.53 \times 10^{-1} \text{ s}$

(d) $0.000274 \text{ kilometers} \times \left(\dfrac{1000 \text{m}}{1 \text{ km}}\right) = 2.74 \times 10^{-4} \text{ m}$

1.53 $45.8 \text{ mg} \times \left(\dfrac{1 \text{oz}}{28350 \text{ mg}}\right) = 1.62 \times 10^{-3} \text{ oz}$

1.55   (a) $(8.2 \times 10^2) + (3.75 \times 10^4) = 820 + 37{,}500 = 3.83 \times 10^4$

(b) $(5.21 \times 10^{-2}) + (2.74 \times 10^{-3}) = 0.0521 + 0.00274 = 5.48 \times 10^{-2}$

(c) $(1.01 \times 10^{-4}) + (7.23 \times 10^{-3}) = 0.000101 + 0.00723 = 7.33 \times 10^{-3}$

1.57 $1.0 \text{ L} \times \left(\dfrac{1000 \text{mL}}{1 \text{ L}}\right) = 1000 \text{ mL} = 1.0 \times 10^3 \text{ cm}^3$

1.59   Heat lost by man = heat gained by water

$$\text{Mass of water} = 1000 \text{ gallons} \times \left(\frac{3.7853\,L}{1\,gal}\right) \times \left(\frac{1000\,mL}{1\,L}\right) \times \left(\frac{1\,g}{1\,mL}\right) = 3{,}785{,}300 \text{ g}$$

$$= C_p \times \text{mass} \times \Delta T = 4.184 \left(\frac{J}{g\,°C}\right) \times 3{,}785{,}300 \text{ g} \times 2.49°C = 3.94 \times 10^7 \text{ J in 1.5 h}$$

$$\left(\frac{3.94 \times 10^7\,J}{1.5\,h}\right) \times 24 \text{ hours} = 6.3 \times 10^8 \text{ J per day.}$$

1.61   The unspecified solution contains two components, NaCl and water. You could obtain pure NaCl by evaporating all the water from the mixture. Five grams of pure NaCl could then easily be weighed out in a laboratory, and the specified mixture of salt and water could be easily prepared.

1.63   You must modify your original hypothesis and retest it.

1.65   The "uk" location stands for United Kingdom (England), a country that uses the SI metric system. In the United States, the Fahrenheit system is still in popular use, so converting 175 Celsius degrees to Fahrenheit degrees should solve the problem.

$$175°C \times (9/5)\,(°F/°C) + 32 = 347°F$$

1.67   (a) 3.60 m × 100 cm/m × 1 in/2.54 cm × 1 ft/12 in × 1 yd/3 ft = 3.94 yd

(b) 26.0 in × 2.54 cm/in = 66.0 cm

(c) 2.99×10^8 m/s × km/1000 m × 1 mile/1.61 km × 60 s/min × 60 min/hr × 24 hr/day × 365 day/yr = 5.90×10^9 miles/yr

(d) This is a fluid measure, so wc use the density of water to convert mL to ounces of water. Assume waters density to 1.00 g/mL.

$$64.0 \text{ mL} \times 1.00 \text{ g/mL} \times 1/28.35 \text{ oz/g} = 2.26 \text{ oz}$$

1.69   The price per mL of the active ingredient is

Recommended product: $6.50/185 mL cream × 1 mL of cream/0.005 mL active ingredient = $7.03/ mL of active ingredient

Generic product: $7.00/165 mL of cream × 1 mL of cream/0.007 mL active ingredient = $6.06/mL of active ingredient

The generic product is more cost effective.

1.71   162/2000 = 8.10%

1.73   Volume of each side of the box = (100 cm × 100 cm × 0.205 cm) = 2.50×10^3 cm^3

Mass of each side of the box = 2.50×10^3 cm^3 × 2.50 g/cm^3 = 6.25×10^3 g

Mass of box = $6.00 \times 6.25 \times 10^3$ = 3.75×10^4 g

Density of box = mass of box/volume of box

$$3.75 \times 10^4 \text{ g}/1.00 \times 10^6 \text{ cm}^3 = 0.0375 \text{ g/cm}^3$$

Since the volume of the box is 1.00×10^6 cm^3, the total amount of water to be added is:

$$3.75 \times 10^4 \text{ mL} + X = 1.00 \times 10^6 \text{ mL}$$

$$X = 9.63 \times 10^5 \text{ mL}$$

1.75   You may have rounded off your figures before you finished the calculation.

Chapter 2

Atomic Structure

2.1   The conservation of mass states that matter cannot be created or destroyed. No mass is lost as a result of a chemical reaction.

2.3 $\dfrac{1.520 \text{ g C}}{3.800 \text{ g}} = 4.000 \times 10^{-1} \times 100 = 40.00\%$

$\dfrac{0.2535 \text{ g H}}{3.800 \text{ g}} = 6.671 \times 10^{-2} \times 100 = 6.671\%$

$\dfrac{2.027 \text{ g O}}{3.800 \text{ g}} = 5.334 \times 10^{-1} \times 100 = 53.33\%$

2.5   Atomic mass of O $= \dfrac{4}{3}\left(12.0000\right) = 16.0000$ amu

2.7   Atomic mass of N $= 0.875 \times 16.0000$ amu $= 14.0$ amu

2.9   Proton: mass 1.00728 amu, charge $= 1+$

Neutron: mass 1.00867 amu, charge $= 0$

Electron: mass 0.0005486 amu, charge $= 1-$

2.11   Atomic mass is the sum of the protons and the neutrons in the atom. In most atoms, the number of protons is close to the number of neutrons and therefore the atomic mass will be approximately double the number of protons, which equals the atomic number of an atom.

2.13   Because the number of electrons is different from the number of protons, it would be a charged ion. Because there are three more protons than electrons, it would be a cation with a charge of 3+.

2.15   Atomic mass of Sr:

$\dfrac{0.5600}{100}\left(83.91\right) + \dfrac{9.860}{100}\left(85.91\right) + \dfrac{7.020}{100}\left(86.91\right) + \dfrac{82.56}{100}\left(87.91\right) = 87.62$ amu

2.17 $^{16}_{8}\text{O}$, $^{17}_{8}\text{O}$; $^{24}_{12}\text{Mg}$, $^{25}_{12}\text{Mg}$; $^{28}_{14}\text{Si}$, $^{29}_{14}\text{Si}$

2.19   (a) lithium, (b) calcium, (c) sodium, (d) phosphorus, (e) chlorine

2.21   Inspection of the periodic table shows that:

Element	Group	Period
Li	I	2
Na	I	3
K	I	4
Rb	I	5
Cs	I	6

2.23   Li, Na, K, Rb, and Cs are metals; they are all on the left side of the periodic table.

2.25   Main-group elements

2.27   Group VII

2.29   Emission occurs when an atom loses energy in going from a high-energy state to a lower one. The ground state is the lowest energy state and therefore the atom cannot undergo a transition to a lower state.

2.31   Shells are identified by a principal quantum number, n. Subshells within shells are identified by the letters $s, p, d, f,$ etc. Orbitals are within subshells.

2.33   An orbital is used to define the probability of finding an electron in a region of space around an atomic nucleus.

2.35   There is one orbital in an s subshell.

2.37   No, an s subshell is spherically symmetrical.

2.39   The number of electrons must equal the number of protons for an element in its standard state. The number of protons identifies the element.

(a) Number of electrons = 12 = number of protons. The element with 12 protons is magnesium.

(b) Number of electrons = 16 = number of protons. The element with 16 protons is sulfur.

(c) Number of electrons = 17 = number of protons. The element with 17 protons is potassium.

2.41   The positive charge indicates that there is one more proton than electron. $1s^2 2s^2 2p^6 = 10$ electrons. Because there must be 11 protons, the cation is Na^+.

2.43   Group I. All elements possess one outer electron, ns^1.

2.45   Group II. All elements possess two outer electrons, ns^2.

2.47   True

2.49   Its energy increases.

2.51   They have virtually identical chemical properties owing to similar outer-shell electronic structure.

2.53   Calcium is a metal; it is found on the left side of the periodic table.

2.55 $^{16}_{8}O$, $^{18}_{8}O$

2.57 $93\ e^- \times \dfrac{1\,amu}{1835\ e^-} = 0.050\ amu$

$\dfrac{0.050\,amu}{237\,amu} = 0.022\%$

2.59   Because energy is proportional to frequency, blue light will have the greater energy.

2.61   The octet rule .

2.63   One needs to know the natural abundances of each of the isotopes.

2.65   Dalton's atom was indestructible, but the "modern" atom can be decomposed into subatomic particles. In addition, the discovery of isotopes showed that the masses of the atoms of an element are not identical.

2.67   Each element has a characteristic spectrum. An elements presence in a sample will be detected by its characteristic spectral lines.

2.69   (a) a discrete bundle of light energy

(b) a number described in a discrete state of an electron

(c) the low-end energy state in an electronic or ion

(d) an allowed energy state that is higher in energy then the ground state

2.71   The repulsion between two similarly charged particles.

2.73   (a) Ca: $[Ar]4s^2$; (b) Br: $[Ar]\ 4s^23d^{10}4p^5$; (c) Ag: $[Kr]5s^14d^{10}$; (d) Zn: $[Ar]4s^23d^{10}$

2.75   (a) $1s^2\ 2s^2\ 2p_x^{\ 1}\ 2p_y^{\ 1}\ 2p_z^{\ 1}$

(b) $1s^2 2s^2 2p_x^1 2p_y^1 2p_z^1$

(c) $1s^2 2s^2 2p_x^2 2p_y^1 2p_z^1$

(d) $1s^2 2s^2 2p_x^2 2p_y^2 2p_z^2$

2.76   (a) two unpaired electrons

(b) zero unpaired electrons

(c) zero unpaired electrons

(d) one unpaired electron

Chapter 3

Molecules and Chemical Bonds

3.1   Chemical bond formation is the result of a process that allows the reacting elements to achieve stability by sharing or transfer of electrons.

3.3   By losing, gaining or sharing electrons.

3.5   These compounds are ionic, thus we do not need to use Greek prefixes (di, tri, …)

(a) K^+ = potassium ion; $(SO_4)^{2-}$ = sulfate ion → K_2SO_4 = potassium sulfate

(b) OH^- = hydroxide ion; Mn^{2+} = Manganese (II) ion → $Mn(OH)_2$ = Manganese (II) hydroxide

(c) NO_3^- = Nitrate ion; Fe^{2+} = iron (II) ion → $Fe(NO_2)_2$ = Iron (II) Nitrate

(d) K^+ = potassium ion; $(H_2PO_4)^-$ = dihydrogen phosphate → KH_2PO_4 = potassium dihydrogen phosphate

(e) Ca^{2+} = Calcium ion; $C_2H_3O_2^-$ = acetate ion → $Ca(C_2H_3O_2)_2$ = calcium acetate

(f) Na^+ = sodium ion; CO_3^{2-} = carbonate ion → Na_2CO_3 = sodium carbonate

3.7   (a) ammonium   (b) nitrate   (c) sulfate   (d) phosphate

(e) acetate

3.9   (a) Aluminum = Al^{3+}; Chloride = Cl^-. Compound must be neutral, therefore $AlCl_3$.

(b) Aluminum = Al^{3+}; Sulfide = S^{2-}. Compound must be neutral, therefore Al_2S_3.

(c) Aluminum = Al^{3+}; Nitride ion = N^{3-}. Compound must be neutral, therefore AlN.

3.11   (a) Lithium = Li^+; oxide = O^{2-}. Compound must be neutral, therefore Li_2O.

(b) Calcium = Ca^{2+}; bromide = Br^-. Compound must be neutral, therefore $CaBr_2$.

(c) Aluminum = Al^{3+}; oxide = O^{2-}. Compound must be neutral, therefore Al_2O_3.

(d) Sodium = Na^+; sulfide = S^{2-} Compound must be neutral, therefore Na_2S.

3.13   Group I forms +1 ions, Group VII form –1 ions → For neutral compound ratio = 1:1

3.15   Group II forms +2 ions, Group VII form –1 ions → For neutral compound ratio = 1:2

3.17   Calcium ion = Ca^{2+}. Any group VI element forms a –2 ion. Therefore to make a neutral compound, 1 calcium and 1 group VI element must be included in the compound. The first three elements in group VI are oxygen, O; sulfur, S; and selenium, Se. Thus,

(a) CaO   (b) CaS   (c) CaSe

3.19   Aluminum ion = Al^{3+}. Any group VI element forms a –2 ion. Therefore to make a neutral compound, 2 aluminums and 3 group VI elements must be included in the compound. The first three elements in group VI are oxygen, O; sulfur, S; and selenium, Se. Thus,

(a) Al_2O_3 (b) Al_2S_3 (c) Al_2Se_3

3.21   Aluminum ion = Al^{3+}. Any group VII element forms a –1 ion. Therefore to make a neutral compound, 1 aluminum and 3 group VII elements must be included in the compound. The first four elements in group VII are fluorine, F; chlorine, Cl; bromine, Br; iodine, I. Thus,

(a) AlF_3 (b) $AlCl_3$ (c) $AlBr_3$ (d) AlI_3

3.23   Procedure to draw Lewis dot structure:

(i)   Place atoms and connect bonds.

(ii)   Determine how many valence electrons are available from all atoms.

(iii) Subtract two electrons for each bond drawn in part (i).

(iv)   Distribute remaining electrons among the atoms. Result should be that every atom has eight electrons associated with it either from free pairs of electrons or covalent bonds.

CCl_4

(i)

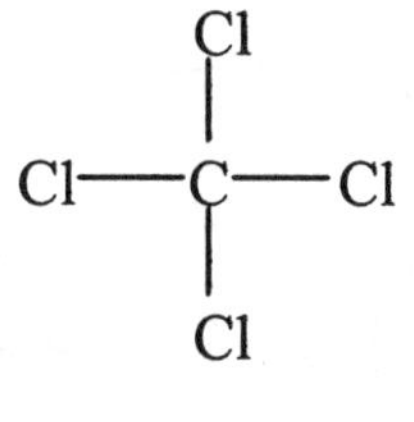

$$
\begin{array}{c}
\text{Cl} \\
| \\
\text{Cl}-\text{C}-\text{Cl} \\
| \\
\text{Cl}
\end{array}
$$

(ii)   4 electrons from the carbon + (7 electrons from each chlorine × 4) = 32 e^-.

(iii) Covalent bonds use 8 electrons → 32 – 8 = 24 electrons remain.

(iv)   Distribute 24 electrons among 4 Cl atoms (6 each) to give each Cl 8 electrons.

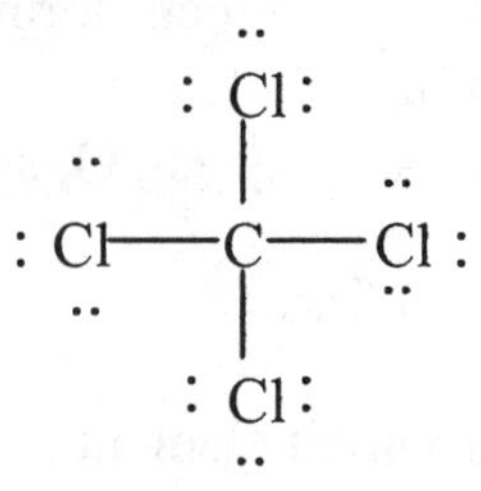

$$\begin{array}{c} \ddot{\;} \\ :\ddot{C}l: \\ | \\ :\ddot{C}l\!-\!C\!-\!\ddot{C}l: \\ | \\ :\ddot{C}l: \\ \ddot{\;} \end{array}$$

Check that all the atoms have 8 electrons associated with it. ---- ✓

3.25   Using (i through iv) in 3.23:

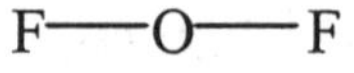

(i)

$$F\!-\!O\!-\!F$$

(ii) 6 electrons from the oxygen + (7 electrons from each fluorine × 2) = 20 e^-.

(iii) Covalent bonds use 4 electrons → 20 – 4 = 16 electrons remain.

(iv) Distribute 16 electrons among F and O. First put 4 electrons (two free pairs) on the oxygen to give it an octet. That leaves 12 electrons, distribute these electrons among 2 F atoms (6 each) to give each F 8 electrons.

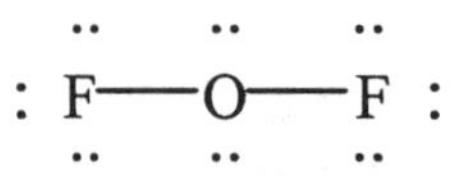

$$:\ddot{F}\!-\!\ddot{O}\!-\!\ddot{F}:$$

Check that all the atoms have eight electrons associated with it. ---- ✓

3.27   Using (i through iv) in 3.23:

(i)

$$O\!-\!C\!-\!O$$

(ii) 4 electrons from the carbon + (6 electrons from each oxygen × 2) = 16 e^-.

(iii) Covalent bonds use 4 electrons → 16 – 4 = 12 electrons remain.

(iv) Distribute 12 electrons among C and O. First put four electrons (two free pairs) on the carbon to give it an octet. That leaves eight electrons, distribute these electrons among two O atoms (four each) to give

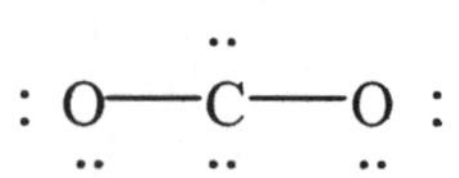

$$:\ddot{O}\!-\!\ddot{C}\!-\!\ddot{O}:$$

Check that all the atoms have eight electrons associated with it. ---- **NO**

Both oxygens have only six electrons. To remedy this, we need to incorporate multiple (double bonds). When there are not enough electrons to fill all atoms to an octet, changing single bonds to double bonds will result in more sharing of electrons.

Try again:

(i) Make bonds between C and O double bonds:

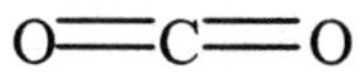

$$O=C=O$$

(ii & iii) Covalent bonds use 8 electrons → 16 – 8 = 8 electrons remain.

(iv) Distribute remaining electrons between the two oxygens as the carbon has eight electrons associated with it.

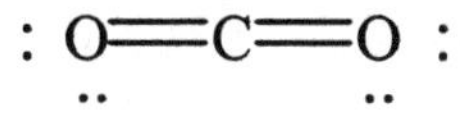

$$: O=C=O :$$

Check that all the atoms have eight electrons associated with it. ---- ✓

3.29   VSEPR provides information on the three dimensional structure of a molecule by taking into account the fact that electrons have a negative charge. Therefore each pair of electrons in an atom wants to be as far away from all other pairs of electrons. In this theory "pairs of electrons" means either non-bonded pairs of electrons or covalent bonds. Furthermore, single, double, and triple bonds all count as a single "pair of electrons".

Therefore, before using VSEPR to determine the three dimensional arrangement of the bonds/non-bonded electron pairs, we must first determine the Lewis dot structures. We will use the procedure outline from Ex. 3.23.

(i)

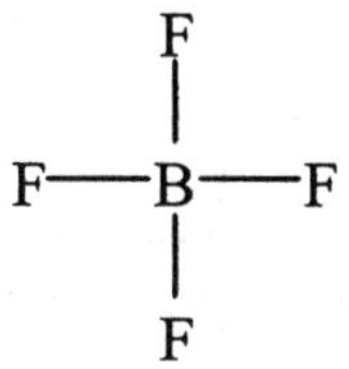

$$\begin{array}{c} F \\ | \\ F - B - F \\ | \\ F \end{array}$$

(ii) 3 electrons from the Boron + (7 electrons from each fluorine × 4) + 1 extra electron from the (–) charge = 32 e^-.

(iii) Covalent bonds use 8 electrons → 32 – 8 = 24 electrons remain.

(iv) Distribute 24 electrons among 4 F (6 each) to give.

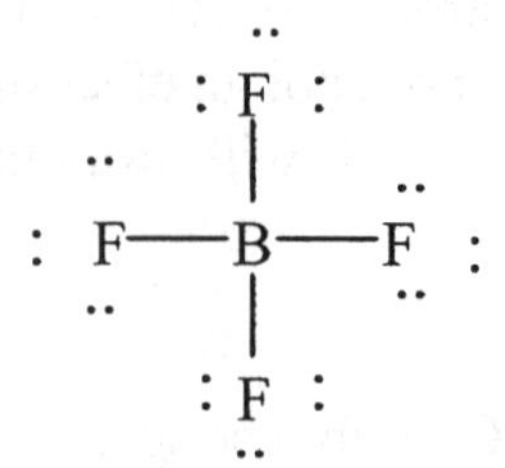

$$
\begin{array}{ccc}
 & \overset{\cdot\cdot}{\underset{}{:\,\ddot{F}\,:}} & \\
 & | & \\
:\ddot{F}\!-\!\!\underset{|}{B}\!\!-\!\ddot{F}: & & \\
 & | & \\
 & \underset{\cdot\cdot}{:\,\ddot{F}\,:} & \\
\end{array}
$$

Therefore, the central atom is the B and it has four bonds, which equals four "pairs of electrons" around it. These four bonds must arrange so that they are as far away from each other as possible. This is accomplished by having the four fluorine atoms situated at the corners of a tetrahedron, with the boron atom at the center.

3.31   The Lewis dot structures are taken from Ex. 3.23, 3.29, and 3.27 for parts a, b, and c, respectively.

(a)

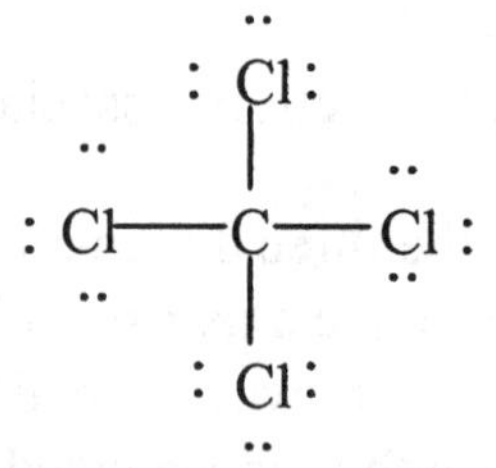

$$
\begin{array}{ccc}
 & \overset{\cdot\cdot}{:\ddot{Cl}:} & \\
 & | & \\
:\ddot{Cl}\!-\!\!\underset{|}{C}\!\!-\!\ddot{Cl}: & & \\
 & | & \\
 & :\ddot{Cl}: & \\
\end{array}
$$

From this structure and VSEPR, CCl_4 forms a tetrahedron with each C–Cl bond going from the center to the corner. Even though each C–Cl bond is polar, this arrangement results in a cancellation of these polar contributions to give a molecule with no dipole moment.

(b)

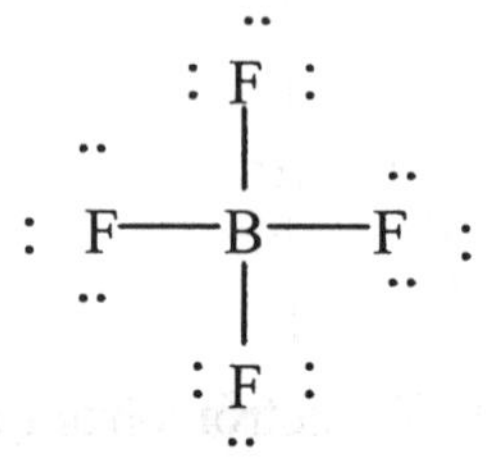

$$
\begin{array}{ccc}
 & \overset{\cdot\cdot}{:\ddot{F}\,:} & \\
 & | & \\
:\ddot{F}\!-\!\!\underset{|}{B}\!\!-\!\ddot{F}: & & \\
 & | & \\
 & :\ddot{F}\,: & \\
\end{array}
$$

From this structure and VSEPR, BF_4 forms a tetrahedron with each B–F bond going from the center to the corner. Even though each B–F bond is polar, this arrangement results in a cancellation of these polar contributions to give a molecule with no dipole moment.

(c)

$$: O = C = O :$$

From this structure and VSEPR, CO_2 forms a linear molecule with each C=O 180° from each other. Even though each C–O– bond is polar, this arrangement results in a cancellation of these polar contributions to give a molecule with no dipole moment.

3.33   A chemical bond that consists of two electrons with paired spins.

3.35   These compounds are ionic, thus we do not need to use Greek prefixes (di, tri, …)

(a) Ag^+ = silver ion; NO_3^- = nitrate ion → $AgNO_3$ = silver nitrate

(b) Hg^{2+} = Mercury (II) ion; Cl^- = chloride ion → $HgCl_2$ = mercury (II) chloride

(c) Na^+ = Sodium ion; CO_3^{2-} = carbonate ion → Na_2CO_3 = sodium carbonate

(d) Ca^{2+} = calcium ion; PO_4^{3-} = phosphate → $Ca_3(PO_4)_2$ = calcium phosphate

3.37   From names of ions above:

(a) magnesium chloride   (b) magnesium sulfide   (c) magnesium nitride

3.39 Ba^{2+} = barium ion; O^{2-} = oxide ion; S^{2-} = sulfide ion; Se^{2-} = selenide ion. Therefore

(a) barium oxide   (b) barium sulfide   (c) barium selenide

3.41 K^+ = potassium ion; N^{3-} = nitride ion; P^{3-} = phosphide ion; As^{3-} = arsenate ion. Therefore:

(a) potassium nitride   (b) potassium phosphide   (c) potassium arsenate

3.43   Te is in the same group as oxygen (group VI) and Br is in the same group as F (group VII). Therefore the Lewis dot structure of $TeBr_2$ will be very similar to OF_2 as determined in 3.25:

$$: Br - Te - Br :$$

Therefore, the central atom is the Te and it has two bonds and two non-bonded pairs, which equals four "pairs of electrons" around it. These four "pairs" must arrange so that they are as far away from each other as possible. This is accomplished by having the two bromine atoms and the two non-bonded pairs situated at the corners of a tetrahedron. Thus, the Br–Te–Br bonds will form a bent molecule.

3.45   Ionic bonds form when the ion attraction power between two atoms is very different (Fig. 3.1) Ion attraction power is related to the difference between the ionization energies of the two elements.

Ionization energies follow the periodic table as shown in Fig. 3.1. Therefore if two atoms are close to each other on the periodic table, they will form covalent bonds, or else they will form ionic bonds

(a) Si and O are close, therefore covalent bonds.

(b) Be and C are close, therefore covalent bonds.

(c) C and N are close, therefore covalent bonds.

(d) Ca and Br are not close, therefore ionic bonds.

(e) B and N are close, therefore covalent bonds.

3.47   Lithium ion = Li^+. Any group VII element forms a –1 ion. Therefore to make a neutral compound, 1 lithium and 1 group VII element must be included in the compound. The first four elements in group VII are fluorine, F; chlorine, Cl; bromine, Br; iodine, I. Thus,

(a) LiF   (b) LiCl   (c) LiBr   (d) LiI

3.49   Yes. Hydrogen only requires two electrons. Also, beryllium and boron require four and six electrons respectively to complete their valence shells.

3.51   Test its ability to conduct electricity when dissolved in water. An ionic substance will dissociate into ions, which will conduct electricity. Thus, if the dissolved substance conducts electricity it is ionic.

3.53   Using (i through iv) in 3.23:

(i)

$$H—O—O—H$$

(ii)   2 electrons from the hydrogens + (6 electrons from each oxygen × 2) = 14 e^-.

(iii)   Covalent bonds use 6 electron → 14 – 6 = 8 electrons remain.

(iv)   Distribute the eight electrons to the oxygen atoms to give:

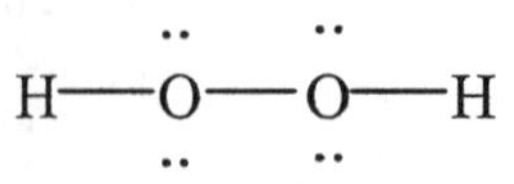

$$H—O—O—H$$

Check that all the atoms have 8 electrons associated with it. ---- ✓

3.55   Using (i through iv) in 3.23:

(i)

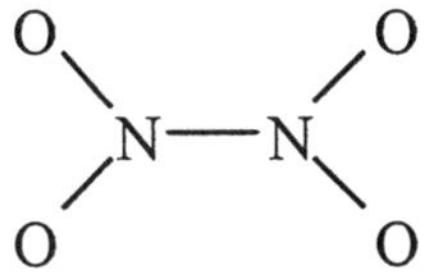

(ii) (5 electrons from each nitrogen × 2) + (6 electrons from each oxygen × 4) = 34 e⁻.

(iii) Covalent bonds use 10 electrons → 34 − 10 = 24 electrons remain.

(iv) Distributing these 24 electrons among the nitrogen and the oxygen atoms to fill the octets does not work. Therefore from Ex. 3.27, make some of the bonds double bonds. When two N–O bonds are made into double bonds, the resultant Lewis dot structure is:

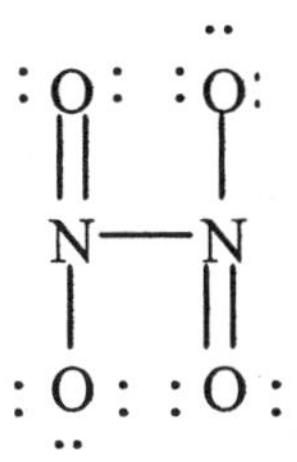

Check that all the atoms have eight electrons associated with it. ---- ✓

3.57   (a) Si-F > Ge-F > C-F

(b) Si-F > Sn-F > C-F

(c) Al-Br > Ga-Br

(d) Sb-Br > As-Br

3.59 (a)

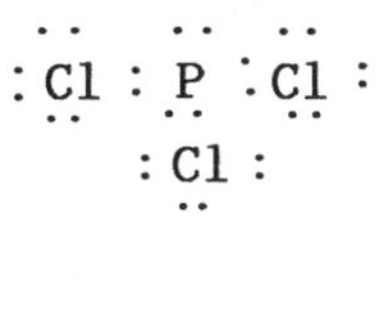

(b)

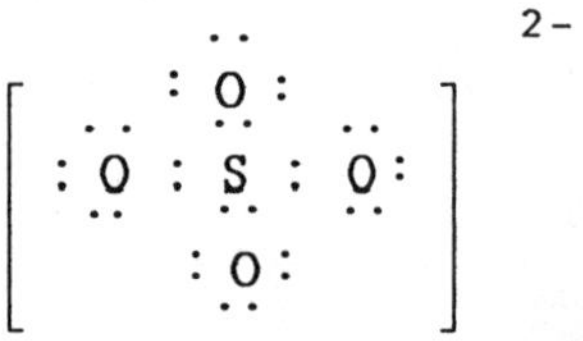

$$\left[\begin{array}{c} \overset{\displaystyle ..}{:}\overset{\displaystyle O}{}: \\ :\overset{..}{\underset{..}{O}} : \overset{..}{\underset{..}{S}} : \overset{..}{\underset{..}{O}}: \\ :\overset{..}{\underset{..}{O}}: \end{array}\right]^{2-}$$

(c)

$$\left[\begin{array}{c} :\overset{..}{O}: \\ :\overset{..}{\underset{..}{O}} : \overset{..}{\underset{..}{P}} : \overset{..}{\underset{..}{O}}: \\ :\overset{..}{\underset{..}{O}}: \end{array}\right]^{3-}$$

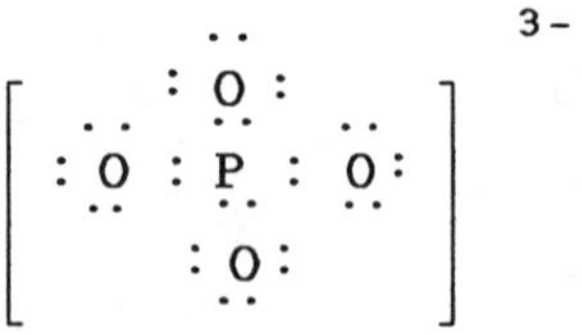

(d)

$$\left[\begin{array}{c} :\overset{..}{O}: \\ :\overset{..}{\underset{..}{O}} : \overset{..}{\underset{..}{Cl}} : \overset{..}{\underset{..}{O}}: \\ :\overset{..}{\underset{..}{O}}: \end{array}\right]^{1-}$$

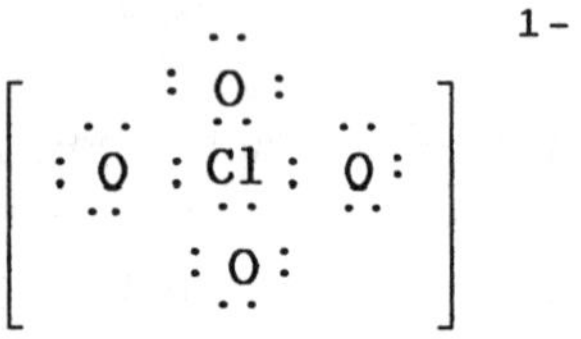

3.61   (a) $[Ar]4s^2 4p^1 + 3\,[He]\,2s^2 2p^5 \longrightarrow [Ar]^{3+} + 3\,[Ne]^{1-}$

(b) $[Kr]4d^{10}5s^1 + [Ne]3s^2 3p^5 \longrightarrow \left[\,[Kr]4d^{10}\,\right]^{1+} + [Ar]^{1-}$

(c) $3\,[He]2s^1 + [He]2s^2 2p^3 \longrightarrow [He]^{1-} + [Ar]^{3-}$

3.63   (a) $TlCl_3$; (b) Zn_2N_3; (c) CdS; (d) GaO

3.65   (a)

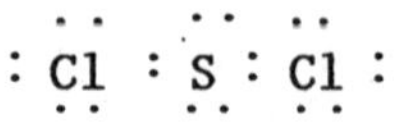

$$:\overset{..}{\underset{..}{Cl}} : \overset{..}{\underset{.}{S}} : \overset{..}{\underset{..}{Cl}}:$$

(b)

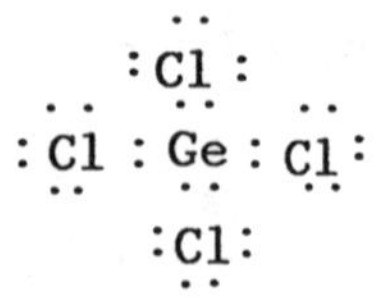

$$\begin{array}{c} :\overset{..}{\underset{}{Cl}}: \\ :\overset{..}{\underset{..}{Cl}} : \overset{}{Ge} : \overset{..}{\underset{..}{Cl}}: \\ :\overset{}{\underset{..}{Cl}}: \end{array}$$

(c)

$$\begin{array}{c} :\overset{..}{\underset{}{Br}} : \overset{.}{As} : \overset{..}{\underset{}{Br}}: \\ :\overset{}{\underset{..}{Br}}: \end{array}$$

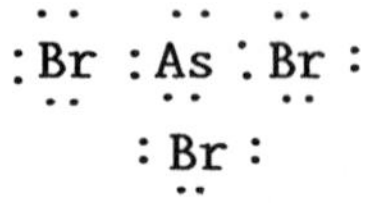

(d)

$$\begin{array}{c} H : \overset{..}{\underset{.}{P}} : H \\ H \end{array}$$

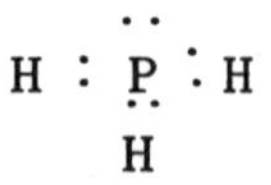

3.67   (a)

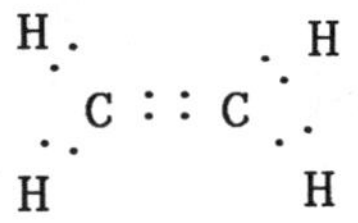

H .   . H
   .   .
     C : : C
   .   .
H .   . H

(b)

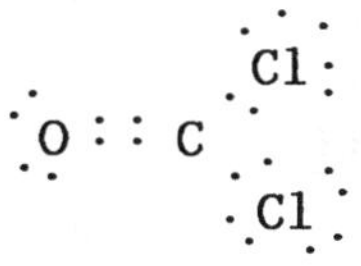

                    . .
               . Cl :
            . .   . .
  . .   .
: O : : C
  . .   .
            . . .
               . Cl .
                    . .

3.69   (a) Linear; (b) Bent; (c) Bent

3.71   (a) Trigonal planar; (b) Bent

Chapter 4

Chemical Calculations

4.1   To determine formula masses of a compound, add the formula masses of all the atoms:

(a) $CaCrO_4$

Atom	Mass	Total
Ca	40.08	40.08
Cr	52.00	52.00
O	16.00	4×16.00
	Total:	156.1

(b) $Mg(OH)_2$

Atom	Mass	Total
Mg	24.31	24.31
H	1.008	2×1.008
O	16.00	2×16.00
	Total:	58.33

(c) $TiCl_4$

Atom	Mass	Total
Ti	47.87	47.87
Cl	35.45	4×35.45
	Total:	189.7

(d) $Na_2Cr_2O_7$

Atom	Mass	Total
Na	22.99	2×22.99
Cr	52.00	2×52.00
O	16.00	7×16.00
	Total:	262.0

(e) C_3H_7OH

Atom	Mass	Total
C	12.01	3×12.01
H	1.008	8×1.008
O	16.00	1×16.00
	Total:	60.09

4.3   To determine formula masses of a compound, add the formula masses of all the atoms:

(a) $Zn_3(AsO_4)_2$

Atom	Mass	Total
Zn	65.39	3×65.41
As	74.92	2×74.92
O	16.00	8×16.00
	Total:	474.1

(b) $Al(C_2H_3O_2)_3$

Atom	Mass	Total
Al	26.98	26.98
C	12.01	6×12.01
H	1.008	9×1.008
O	16.00	6×16.00
	Total:	204.1

(c) $HgCl_2$

Atom	Mass	Total
Hg	200.59	1×200.59
Cl	35.45	2×35.45
	Total:	271.5

(d) $Sr(ClO_3)_2$

Atom	Mass	Total
Sr	87.62	87.62
Cl	35.45	2×35.45
O	16.00	6×16.00
	Total:	254.5

(e) BI_3

Atom	Mass	Total
B	10.81	1×10.81
I	126.9	3×126.9
	Total:	391.5

4.5   To convert from the number of grams to moles, the number of grams must be divided by the formula weight:

$$24.67 \text{ g NaCl} \times \frac{1 \text{ mole NaCl}}{58.44 \text{ g NaCl}} = 0.4221 \text{ mol NaCl}$$

4.7   To convert from the number of moles to grams, the number of moles must be multiplied by the formula weight:

$$0.275 \text{ mol Na}_5\text{P}_3\text{O}_{10} \times \frac{367.86 \text{ g Na}_5\text{P}_3\text{O}_{10}}{1 \text{ mol Na}_5\text{P}_3\text{O}_{10}} = 101 \text{ g Na}_5\text{P}_3\text{O}_{10}$$

4.9   To convert from moles to the number of molecules, the number of moles must be multiplied by Avogadro's number, 6.022×10^{23} molecules/mol. In this problem, one must take into account the fact that each molecule creates two Zn ions.

$$30.5 \text{ g Zn}_2\text{P}_2\text{O}_7 \times \frac{1 \text{ mol}}{304.7 \text{ g}} \times \frac{2 \text{ mol Zn ions}}{1 \text{ mol Zn}_2\text{P}_2\text{O}_7} \times \frac{6.022 \times 10^{23} \text{ ion}}{1 \text{ mol Zn ions}} = 1.21 \times 10^{23} \text{ Zn ions}$$

4.11   To convert from the number of molecules to moles, the number of molecules must be divided by Avogadro's number, 6.022×10^{23} molecules/mol.

$$10.54 \times 10^{24} \text{ molecules} \times \frac{1 \text{ mol}}{6.022 \times 10^{23} \text{ molecules}} \times \frac{16.04 \text{ g CH}_4}{1 \text{ mol CH}_4} = 280.7 \text{ g CH}_4$$

4.13   Apply these proportions to a sample that is 100 grams:

$$36.05 \text{ g Zn} \times \frac{1 \text{ mol}}{65.38 \text{ g}} = 0.5513 \text{ mol Zn}$$

$$28.67 \text{ g Cr} \times \frac{1 \text{ mol}}{52.00 \text{ g}} = 0.5513 \text{ mol Cr}$$

$$35.28 \text{ g O} \times \frac{1 \text{ mol}}{16.00 \text{ g}} = 2.205 \text{ mol O}$$

Divide each molar amount by the smallest value to get whole numbers:

$0.5513/0.5513 = 1$ $\qquad$ $0.5513/0.5513 = 1$ $\qquad$ $2.205/0.5513 = 3.999 \approx 4$

Thus, the empirical formula $ZnCrO_4$

4.15   (a) To balance equation, concentrate on one atom at a time.

    (i) Since there are 2 N atoms on the lefthand side of the equation and only 1 N atom on the righthand side of the equation, put a coefficient of 2 in front of NBr_3 to equate the number of N atoms on both sides of the reaction equation:

$$N_2 + Br_2 \rightarrow 2 \, NBr_3$$

    (ii) Now, since there are 6 Br atoms on the righthand side and only 2 Br atoms on the lefthand side, put a coefficient of 3 in front of Br_2 to equate the number of Br atoms on both sides of the reaction equation:

$$N_2 + 3 \, Br_2 \rightarrow 2 \, NBr_3$$

(b) $2 \, HNO_3 + Ba(OH)_2 \rightarrow Ba(NO_3)_2 + 2 \, H_2O$

(c) $HgCl_2 + H_2S \rightarrow HgS + 2 \, HCl$

4.17   (a) $4 \, P + 5 \, O_2 \rightarrow 2 \, P_2O_5$

(b) $FeCl_2 + K_2SO_4 \rightarrow FeSO_4 + 2 \, KCl$

(c) $HgCl + 2 \, NaOH + 2 \, NH_4Cl \rightarrow Hg(NH_3)_2Cl + 2 \, NaCl + 2 \, H_2O$

4.19   Hydrocarbons combust to form CO_2 and H_2O:

$$C_6H_6 + O_2 \rightarrow CO_2 + H_2O$$

Now balance as in examples 4.15 through 4.18.

(i) Since there are 6 C atoms on the lefthand side of the equation and only 1 C atom on the righthand side of the equation, put a coefficient of 6 in front of CO_2 to equate the number of C atoms on both sides of the reaction equation:

$$C_6H_6 + O_2 \rightarrow 6\ CO_2 + H_2O$$

(ii) Now, since there are 6 H atoms on the lefthand side and only 2 H atoms on the righthand side, put a coefficient of 3 in front of H_2O to equate the number of H atoms on both sides of the reaction equation:

$$C_6H_6 + O_2 \rightarrow 6\ CO_2 + 3\ H_2O$$

(ii) Now, there are 2 O atoms on the lefthand side and 15 O atoms on the righthand side; the simplest way to equate the number of O atoms on both sides of the equation is to place a stoichiometric prefactor of $\dfrac{15}{2}$ in front of O_2:

$$C_6H_6 + \frac{15}{2}\ O_2 \rightarrow 6\ CO_2 + 3\ H_2O$$

(ii) Multiply through by 2 to obtain integer coefficients:

$$2\ C_6H_6 + 15\ O_2 \rightarrow 12\ CO_2 + 6\ H_2O$$

4.21 $C_5H_{12} + 8\ O_2 \rightarrow 5\ CO_2 + 6\ H_2O$

$2\ C_8H_{18} + 25\ O_2 \rightarrow 16\ CO_2 + 18\ H_2O$

$C_{10}H_{20} + 15\ O_2 \rightarrow 10\ CO_2 + 10\ H_2O$

4.23 $CH_4 + 2\ O_2 \rightarrow CO_2 + 2\ H_2O$

4.25   Unit conversion factors are ratios of amount of one component to amount of another component, in moles:

$$\frac{2\ Al(OH)_3}{3\ H_2SO_4}, \frac{1\ Al(OH)_3}{3\ H_2O}, \frac{1\ H_2SO_4}{2\ H_2O}, \frac{2\ Al(OH)_3}{1\ Al_2(SO_4)_3}, \frac{3\ H_2SO_4}{1\ Al_2(SO_4)_3}, \frac{1\ Al_2(SO_4)_3}{6\ H_2O}$$

and the inverse of each of these.

4.27   From problem 4.25:

$$2\ Al(OH)_3 + 3\ H_2SO_4 \rightarrow Al_2(SO_4)_3 + 6\ H_2O$$

$$2.5 \text{ mol H}_2\text{SO}_4 \times \frac{1 \text{ mol Al}_2(\text{SO}_4)_3}{3 \text{ mol H}_2\text{SO}_4} = 0.83 \text{ mol Al}_2(\text{SO}_4)_3$$

4.29 $$6.5 \text{ mol H}_2\text{O} \times \frac{3 \text{ mol H}_2\text{SO}_4}{6 \text{ mol H}_2\text{O}} = 3.3 \text{ mol H}_2\text{SO}_4$$

4.31   Note: Information is in grams, answer is in grams, must still do conversions in moles:

$$46.0 \text{ g H}_2\text{SO}_4 \times \frac{1 \text{ mol H}_2\text{SO}_4}{98.0 \text{ g H}_2\text{SO}_4} = 0.469 \text{ mol H}_2\text{SO}_4$$

$$0.469 \text{ mol H}_2\text{SO}_4 \times \frac{1 \text{ mol Al}_2(\text{SO}_4)_3}{3 \text{ mol H}_2\text{SO}_4} = 0.156 \text{ mol Al}_2(\text{SO}_4)_3$$

$$0.156 \text{ mol Al}_2(\text{SO}_4)_3 \times \frac{342.2 \text{ g Al}_2(\text{SO}_4)_3}{1 \text{ mol Al}_2(\text{SO}_4)_3} = 53.4 \text{ g Al}_2(\text{SO}_4)_3$$

4.33 $2 \text{ Cu(NO}_3)_2 + 4 \text{ NaI} \rightarrow \text{I}_2 + 2 \text{ CuI} + 4 \text{ NaNO}_3$

(a) $$0.87 \text{ mol NaI} \times \frac{2 \text{ mol Cu(NO}_3)_2}{4 \text{ mol NaI}} = 0.44 \text{ mol Cu(NO}_3)_2$$

(b) $$1.45 \text{ mol CuI} \times \frac{4 \text{ mol NaI}}{2 \text{ mol CuI}} = 2.90 \text{ mol NaI}$$

(c) $$15.0 \text{ g NaI} \times \frac{1 \text{ mol NaI}}{149.89 \text{ g NaI}} = 0.100 \text{ mol NaI}$$

$$0.100 \text{ mol NaI} \times \frac{2 \text{ mol Cu(NO}_3)_2}{4 \text{ mol NaI}} = 0.050 \text{ mol Cu(NO}_3)_2$$

$$0.050 \text{ mol Cu(NO}_3)_2 \times \frac{187.56 \text{ g Cu(NO}_3)_2}{1 \text{ mol Cu(NO}_3)_2} = 9.4 \text{ g Cu(NO}_3)_2$$

(d) $$15.00 \text{ g NaI} \times \frac{1 \text{ mol NaI}}{149.89 \text{ g NaI}} = 0.1000 \text{ mol NaI}$$

$$0.1000 \text{ mol NaI} \times \frac{4 \text{ mol NaNO}_3}{4 \text{ mol NaI}} = 0.1000 \text{ mol NaNO}_3$$

$$0.1000 \text{ mol NaNO}_3 \times \frac{85.00 \text{ g NaNO}_3}{1 \text{ mol NaNO}_3} = 8.500 \text{ g NaNO}_3$$

4.35   (a) $\text{Ca}_3(\text{PO}_4)_2 + 4 \text{ H}_3\text{PO}_4 \rightarrow 3 \text{ Ca(H}_2\text{PO}_4)_2$

(b) $FeCl_2 + (NH_4)_2S \rightarrow FeS + 2 NH_4Cl$

(c) $2 KClO_3 \rightarrow 2 KCl + 3 O_2$

(d) $3 O_2 \rightarrow 2 O_3$

(e) $2 C_5H_6 + 13 O_2 \rightarrow 10 CO_2 + 6 H_2O$

4.37 (a) $102.6 \text{ g BaCO}_3 \times \dfrac{1 \text{ mol BaCO}_3}{197.3 \text{ g BaCO}_3} = 0.5200 \text{ mol BaCO}_3$

(b) $60.75 \text{ g HBr} \times \dfrac{1 \text{ mol HBr}}{80.92 \text{ g HBr}} = 0.7507 \text{ mol HBr}$

(c) $148.5 \text{ g CuCl} \times \dfrac{1 \text{ mol CuCl}}{99.00 \text{ g CuCl}} = 1.500 \text{ mol CuCl}$

(d) $50.4 \text{ g HNO}_3 \times \dfrac{1 \text{ mol HNO}_3}{63.02 \text{ g HNO}_3} = 0.800 \text{ mol HNO}_3$

(e) $65 \text{ g C}_6\text{H}_{12}\text{O}_6 \times \dfrac{1 \text{ mol C}_6\text{H}_{12}\text{O}_6}{180.2 \text{ g C}_6\text{H}_{12}\text{O}_6} = 0.36 \text{ mol C}_6\text{H}_{12}\text{O}_6$

4.39 The formula weight for $Ca(OH)_2$ is 74.07 grams/mole.

$$133 \text{ g Ca(OH)}_2 \times \dfrac{1 \text{ formula unit Ca(OH)}_2}{74.1 \text{ formula weight Ca(OH)}_2} = 1.80 \text{ formula units Ca(OH)}_2$$

4.41 When sodium is oxidized, each atom loses 1 electron:

$$Na \rightarrow Na^+ + 1 e^-$$

$$42 \text{ g Na} \times \dfrac{1 \text{ mol Na}}{22.99 \text{ g Na}} \times \dfrac{6.022 \times 10^{23} \text{ Na atoms}}{1 \text{ mol Na}} \times \dfrac{1 e^-}{1 \text{ Na atom}} = 1.1 \times 10^{24} \text{ electrons}$$

4.43 Assume 100 g sample of silver chloride:

$$75.26 \text{ g Ag} \times \dfrac{1 \text{ mol Ag}}{107.87 \text{ g Ag}} = 0.6977 \text{ mol Ag}$$

$$24.74 \text{ g Cl} \times \dfrac{1 \text{ mol Cl}}{35.45 \text{ g Cl}} = 0.6978 \text{ mol Cl}$$

Dividing by the smaller number of relative moles: for both Ag and Cl, 0.6977/0.6978 = 1. Therefore, the empirical formula is AgCl.

4.45   Assume 100 g sample of the methyl ether:

$$52.17 \text{ g C} \times \frac{1 \text{ mol C}}{12.01 \text{ g C}} = 4.344 \text{ mol C}$$

$$13.05 \text{ g H} \times \frac{1 \text{ mol H}}{1.008 \text{ g H}} = 12.95 \text{ mol H}$$

$$34.78 \text{ g O} \times \frac{1 \text{ mol O}}{16.00 \text{ g O}} = 2.174 \text{ mol O}$$

Dividing by the smaller number of relative moles: for C, 4.344/2.174 = 2; for H, 12.95/2.17 = 6; for O, 2.174/2.174 = 1. Therefore, the empirical formula is C_2H_6O.

4.47   (a) $AgNO_3 + KCl \rightarrow AgCl + KNO_3$

(b) $Ba(NO_3)_2 + Na_2SO_4 \rightarrow BaSO_4 + 2 NaNO_3$

(c) $2 (NH_4)_3PO_4 + 3 Ca(NO_3)_2 \rightarrow Ca_3(PO_4)_2 + 6 NH_4NO_3$

(d) $3 Mg(OH)_2 + 2 H_3PO_4 \rightarrow Mg_3(PO_4)_2 + 6 H_2O$

4.49 $2.000 \text{ mol Al}_2(SO_4)_3 \times \dfrac{342.2 \text{ g Al}_2(SO_4)_3}{1 \text{ mol Al}_2(SO_4)_3} = 684.4 \text{ g Al}_2(SO_4)_3$

4.51   Assume 1 mol of compound:

$$17 \text{ mol C} \times \frac{12.01 \text{ g C}}{1 \text{ mol C}} = 204.2 \text{ g C} \qquad \% \text{ C}: \frac{204.2 \text{ g C}}{303.4 \text{ g total}} = 67.3 \% \text{ C}$$

$$21 \text{ mol H} \times \frac{1.01 \text{ g H}}{1 \text{ mol H}} = 21.2 \text{ g H} \qquad \% \text{ H}: \frac{21.2 \text{ g H}}{303.4 \text{ g total}} = 7.0 \% \text{ H}$$

$$1 \text{ mol N} \times \frac{14.0 \text{ g N}}{1 \text{ mol N}} = 14.0 \text{ g N} \qquad \% \text{ N}: \frac{14.0 \text{ g N}}{303.4 \text{ g total}} = 4.6 \% \text{ N}$$

$$4 \text{ mol O} \times \frac{16.0 \text{ g O}}{1 \text{ mol O}} = 64.0 \text{ g O} \qquad \% \text{ O}: \frac{64.0 \text{ g O}}{303.4 \text{ g total}} = 21.1 \% \text{ O}$$

4.53   (a) $0.46 \text{ mol H}_2SO_4 \times \dfrac{98.18 \text{ g H}_2SO_4}{1 \text{ mol H}_2SO_4} = 45 \text{ g H}_2SO_4$

(b) $1.80 \text{ mol Ca(OH)}_2 \times \dfrac{74.1 \text{ g Ca(OH)}_2}{1 \text{ mol Ca(OH)}_2} = 133 \text{ g Ca(OH)}_2$

(c) $0.76 \text{ mol C}_2\text{H}_6\text{O} \times \dfrac{46.07 \text{ g C}_2\text{H}_6\text{O}}{1 \text{ mol C}_2\text{H}_6\text{O}} = 35 \text{ g C}_2\text{H}_6\text{O}$

(d) $1.89 \text{ mol C}_4\text{H}_{10} \times \dfrac{58.1 \text{ g C}_4\text{H}_{10}}{1 \text{ mol C}_4\text{H}_{10}} = 110 \text{ g C}_4\text{H}_{10}$

(e) $1.66 \text{ mol FeCl}_2 \times \dfrac{126.8 \text{ g FeCl}_2}{1 \text{ mol FeCl}_2} = 211 \text{ g FeCl}_2$

4.55   Assume 100 g sample of the oxide of phosphorus:

$43.64 \text{ g P} \times \dfrac{1 \text{ mol P}}{30.97 \text{ g P}} = 1.409 \text{ mol P}$

$56.36 \text{ g O} \times \dfrac{1 \text{ mol O}}{16.00 \text{ g O}} = 3.522 \text{ mol O}$

Dividing by the smaller number of relative moles: for P, 1.409/1.409 = 1; for O, 3.522/1.409 = 2.5. These are not all integers; therefore, they must be multiplied by a factor (2) to convert them to integers to give the empirical formula P_2O_5.

The molecular mass was shown to be 282 g/mol. This is two times the empirical weight of 141 g/mol. Thus, the molecular formula is twice that calculated from the weight analysis: P_4O_{10}.

4.57   (a) $75.0 \text{ g Al}_2(\text{SO}_4)_3 \times \dfrac{1 \text{ mol Al}_2(\text{SO}_4)_3}{342.1 \text{ g Al}_2(\text{SO}_4)_3} = 0.219 \text{ mol Al}_2(\text{SO}_4)_3$

(b) $25.0 \text{ g K}_2\text{Cr}_2\text{O}_7 \times \dfrac{1 \text{ mol K}_2\text{Cr}_2\text{O}_7}{294.2 \text{ g K}_2\text{Cr}_2\text{O}_7} = 0.0850 \text{ mol K}_2\text{Cr}_2\text{O}_7$

4.59 $\dfrac{6.022 \times 10^{23} \text{ dollars}}{6.4 \times 10^9 \text{ people}} = \$9.4 \times 10^{13} \text{ per person}$

4.61   (a) $\dfrac{342.3 \text{ g C}_{12}\text{H}_{22}\text{O}_{11}}{1 \text{ mol C}_{12}\text{H}_{22}\text{O}_{11}} \times \dfrac{1 \text{ mol C}_{12}\text{H}_{22}\text{O}_{11}}{6.022 \times 10^{23} \text{ molecules C}_{12}\text{H}_{22}\text{O}_{11}} = 5.684 \times 10^{-22} \text{ g}$

(b)
$\dfrac{151.9 \text{ g FeSO}_4}{1 \text{ mol FeSO}_4} \times \dfrac{1 \text{ mol FeSO}_4}{6.022 \times 10^{23} \text{ molecules FeSO}_4} \times \dfrac{1 \text{ formula unit FeSO}_4}{1 \text{ mol FeSO}_4} = 2.52 \times 10^{-22} \text{ g}$

4.63   10.0 g cobalt = 0.170 moles of cobalt. This reacts with 12.0 (22.0 − 10.0) grams of chlorine to form the final product. 12.00 g Cl = 0.339 moles of Cl. Therefore, there are twice as many moles of chlorine atoms in the final compound than cobalt, thus the compound is $CoCl_2$.

4.65   The new compound added $(2.166 - 2.006)$ g of oxygen upon reaction $= 0.160$ g $= 0.0100$ moles of oxygen. Since the molecular formula is MO, there are as many moles of metal in the final compound as moles of oxygen. Therefore,

$$2.006 \text{ grams of metal} = 0.0100 \text{ mol of metal}$$

$$2.006 \text{ g} / 0.0100 \text{ mol} = 200.6 \text{ g/mol} = 200.6 \text{ amu}$$

4.67   One molecule of Fe has a mass of 9.26×10^{-23} g

$$\frac{55.85 \text{ g Fe}}{1 \text{ mol Fe}} \times \frac{1 \text{ mol Fe}}{6.022 \times 10^{23} \text{ molecules Fe}} = 9.27 \times 10^{-23} \text{ g}$$

9.27×10^{-23} g is 0.349% of total mass of the myoglobin molecule, X

$$9.27 \times 10^{-23} \text{ g} = 0.00349 \times X$$

$$X = 2.66 \times 10^{-20} \text{ g}$$

$$\frac{2.66 \times 10^{-20} \text{ g}}{\text{myoglobin molecule}} \times \frac{6.022 \times 10^{23} \text{ molecules myoglobin}}{1 \text{ mol myoglobin}} = \frac{1.60 \times 10^4 \text{ g myoglobin}}{1 \text{ mol myoglobin}}$$

Thus, the molecular mass of myoglobin is 1.60×10^4 amu

4.69   (a) $3 \text{ NaHCO}_3 + \text{C}_6\text{H}_8\text{O}_7 \rightarrow 3 \text{ CO}_2 + 3 \text{ H}_2\text{O} + \text{Na}_3\text{C}_6\text{H}_5\text{O}_7$

(b)
$$0.200 \text{ g NaHCO}_3 \times \frac{1 \text{ mol NaHCO}_3}{83.9 \text{ g NaHCO}_3} \times \frac{1 \text{ mol C}_6\text{H}_8\text{O}_7}{3 \text{ mol NaHCO}_3} \times \frac{192 \text{ g C}_6\text{H}_8\text{O}_7}{1 \text{ mol C}_6\text{H}_8\text{O}_7} = 0.152 \text{ g C}_6\text{H}_8\text{O}_7$$

(c) $0.200 \text{ g NaHCO}_3 \times \dfrac{1 \text{ mol NaHCO}_3}{83.9 \text{ g NaHCO}_3} \times \dfrac{3 \text{ mol CO}_2}{3 \text{ mol NaHCO}_3} \times \dfrac{44.0 \text{ g CO}_2}{1 \text{ mol CO}_2} = 0.105 \text{ g CO}_2$

4.71   Note that this reaction takes place in an acidic solution. The first step is to separate the oxidation half-reaction from the reduction half-reaction:

Step 1. $Fe^{2+} \longrightarrow Fe^{3+}$ oxidation

          $MnO_4^- \longrightarrow Mn^{2+}$ reduction

Step 2.   Make certain that both equations are balanced with respect to all elements other than oxygen and hydrogen. In this case, they are balanced.

Step 3.   Balance the oxygen atoms. This is done by noting that the oxygen of MnO^{4-} must appear as H_2O, one of the products of the reduction step. In this case, only the manganese half-reaction involves oxygen. There are four oxygen atoms on the left, none on the right, so we balance the oxygen atoms by adding 4 H_2O to the right-hand side of the manganese half-reaction:

$$Fe^{2+} \longrightarrow Fe^{3+}$$

$$MnO_4^- \longrightarrow Mn^{2+} + 4 H_2O$$

Step 4.   Balance with respect to hydrogen. There are eight hydrogen atoms on the righthand side of the MnO_4^- half-reaction, so we add 8 H^+ ions to the left-hand side of that half-reaction:

$$Fe^{2+} \longrightarrow Fe^{3+}$$

$$8 H^+ + MnO_4^- \longrightarrow Mn^{2+} + 4 H_2O$$

The two half-reactions are now balanced with respect to atoms, but not with respect to charge.

Step 5.   Now balance each half-reaction with respect to charge by adding electrons to the side with excess positive charge. The iron half-reaction has a charge of +2 on the left, and +3 on the right, so we add one electron to the right-hand side:

$$Fe^{2+} \longrightarrow Fe^{3+} + e^-$$

The manganese half-reaction has a net charge of +7 on the left-hand side, and +2 on the right-hand side, so we add five electrons to the left-hand side.

$$5e^- + 8 H^+ + MnO_4^- \longrightarrow Mn^{2+} + 4 H_2O$$

The situation at this stage is that both equations are completely balanced with respect to atoms and charge. It is now necessary to recognize that there must be equal exchanges of electrons in a redox reaction; numbers of electrons consumed must be equal to the numbers of electrons produced. The oxidation of one mole of Fe^{2+} produces one mole of electrons. The reduction of one mole of MnO_4^- requires five moles of electrons. This is accounted for by multiplying the iron half-reaction by five:

$$5 Fe^{2+} \longrightarrow 5 Fe^{3+} + 5 e^-$$

$$5e^- + 8 H^+ + MnO_4^- \longrightarrow Mn^{2+} + 4 H_2O$$

Step 6.   Obtain the complete balanced equation by adding the two half-reactions and canceling any terms that appear on both sides:

$$5 \ Fe^{2+} \longrightarrow 5 \ Fe^{3+} + 5 \ e^{-}$$

$$\underline{5e^{-} + 8 \ H^{+} + MnO_4^{-} \longrightarrow Mn^{2+} + 4 \ H_2O}$$

$$5 \ Fe^{2+} + MnO_4^{-} + 8 \ H^{+} \longrightarrow 5 \ Fe^{3+} + Mn^{2+} + 4 \ H_2O$$

In the finished equation, the electrons do not appear. They were equal on both sides of the equation, and therefore canceled. Properly done, electrons will never appear in the complete balanced equation. The last step is to rewrite the complete equation indicating appropriate physical states:

$$5 \ Fe^{2+} \ (aq) + MnO_4^{-} \ (aq) + 8 \ H^{+}(aq) \longrightarrow 5 \ Fe^{3+} \ (aq) + Mn^{2+}(aq) + 4 \ H_2O(l)$$

4.73   <u>Oxidized</u>   <u>Reduced</u>

(a) I_2 I_2

(b)   Al OH^{-}

Chapter 5

The Physical Properties of Gases

5.1   (a) $364 \text{ mm Hg} \times \dfrac{1 \text{ torr}}{1 \text{ mm Hg}} = 364 \text{ torr.}$

       (b) $483 \text{ mm Hg} \times \dfrac{1 \text{ torr}}{1 \text{ mm Hg}} = 483 \text{ torr.}$

       (c) $675 \text{ mm Hg} \times \dfrac{1 \text{ torr}}{1 \text{ mm Hg}} = 675 \text{ torr.}$

       (d) $735 \text{ mm Hg} \times \dfrac{1 \text{ torr}}{1 \text{ mm Hg}} = 735 \text{ torr.}$

5.3   (a) $735 \text{ mm Hg} \times \dfrac{1 \text{ torr}}{1 \text{ mm Hg}} = 735 \text{ torr.}$

       (b) $735 \text{ mm Hg} \times \dfrac{1 \text{ atm}}{760 \text{ mm Hg}} = 0.967 \text{ atm.}$

5.5   (a) $1.20 \text{ atm} \times \dfrac{760 \text{ torr}}{1 \text{ atm}} = 912 \text{ torr.}$

       (b) $0.450 \text{ atm} \times \dfrac{760 \text{ mm Hg}}{1 \text{ atm}} = 342 \text{ mm Hg.}$

       (c) $850 \text{ mm Hg} \times \dfrac{1 \text{ atm}}{760 \text{ mm Hg}} = 1.12 \text{ atm.}$

5.7   Temperature, T, and the number of moles, n.

5.9   Volume, V, and the number of moles, n.

5.11 $P_1V_1 = P_2V_2$

       $(1.0 \text{ atm}) (2.0 \text{ L}) = (0.80 \text{ atm}) (V_2) \;\rightarrow\; V_2 = 2.5 \text{ L}$

5.13 $\dfrac{P_1V_1}{T_1} = \dfrac{P_2V_2}{T_2} \;\rightarrow\; \dfrac{(1.00 \text{ L})(0.800 \text{ atm})}{298 \text{ K}} = \dfrac{(1.00 \text{ atm})V_2}{373 \text{ K}}$

       $V_2 = 1.00 \text{ L}$

5.15 $\dfrac{P_1V_1}{T_1} = \dfrac{P_2V_2}{T_2} \rightarrow \dfrac{(1.20\ L)(0.800\ atm)}{298\ K} = \dfrac{(1.00\ atm)(2.00\ L)}{T_2}$

$T_2 = 621\ K = 348\ °C$

5.17 $\dfrac{P_1V_1}{T_1} = \dfrac{P_2V_2}{T_2} \rightarrow \dfrac{(1.50\ atm)(1.20\ L)}{298\ K} = \dfrac{P_2(0.600\ L)}{323\ K}$

$P_2 = 3.25\ atm.$

5.19 $\dfrac{P_1V_1}{T_1} = \dfrac{P_2V_2}{T_2} \rightarrow P_2 = 2P_1;\ T_2 = 1.5T_1$

$\dfrac{P_1V_1}{T_1} = \dfrac{2P_1V_2}{1.5T_1} \rightarrow V_2 = \dfrac{P_1}{2P_1}\dfrac{1.5T_1}{T_1}V_1$

$V_2 = 0.750 \times V_1 = 2.25\ L$

5.21   At STP, P = 1.00 atm and T = 273 K,

$\dfrac{P_1V_1}{T_1} = \dfrac{P_2V_2}{T_2} \rightarrow \dfrac{(0.830\ atm)(2.00\ L)}{298\ K} = \dfrac{(1.00\ atm)V_2}{273\ K}$

$V_2 = 1.52\ L$

5.23   At STP, P = 1.00 atm; T = 273 K; 1 mole of gas occupies 22.4 L.

$P = 1.00\ atm \times \dfrac{760\ torr}{1\ atm} = 760\ torr;\ T = 273\ K;\ n = 1\ mol;$

$V = 22.4\ L \times \dfrac{1000\ ml}{L} = 2.25 \times 10^4\ ml$

$R = \dfrac{PV}{nT} = \dfrac{(760\ torr)(2.24 \times 10^4\ ml)}{(1\ mol)(273\ K)} = 6.24 \times 10^4\ \dfrac{torr \times ml}{mol \times K}$

5.25   PV=nRT; $n_{final} = n_{initial} + 0.100\ mol$

$n_{initial} = \dfrac{PV}{RT} = \dfrac{(1\ atm)(1.35\ L)}{(273\ K)(0.0821\ \dfrac{L \times atm}{K \times mol})} = 0.0600\ mol$

$n_{final} = 0.160\ mol$

$V = \dfrac{nRT}{P} = \dfrac{(0.160\ mol)(0.0821\dfrac{L \cdot atm}{K \cdot mol})(273\ K)}{1\ atm} = 3.59\ L$

5.27 $8.00 \text{ g O}_2 \times \dfrac{1.0 \text{ mol O}_2}{32.0 \text{ g O}_2} = 0.250 \text{ mol O}_2$

$$V = \frac{nRT}{P} = \frac{(0.250 \text{ mol})(0.0821 \frac{L \cdot atm}{K \cdot mol})(273 \text{ K})}{1 \text{ atm}} = 5.60 \text{ L}$$

5.29 $P = 1 \text{ atm}; \; T = 273 \text{ K}; \; R = 0.0821 \dfrac{L \cdot atm}{mol \cdot K}$

$V(NH_3, \text{ ammonia}) = 5.60 \text{ L}; \;\; V(H_2, \text{ hydrogen}) = 11.2 \text{ L}$

ammonia → $n = \dfrac{PV}{RT} = \dfrac{(1 \text{ atm})(5.60 \text{ L})}{(273 \text{ K})(0.0821 \frac{L \cdot atm}{K \cdot mol})} = 0.250 \text{ mole} \times \dfrac{17.0 \text{ g NH}_3}{1 \text{ mol NH}_3} = 4.25 \text{ g}$

hydrogen → $n = \dfrac{PV}{RT} = \dfrac{(1 \text{ atm})(11.2 \text{ L})}{(273 \text{ K})(0.0821 \frac{L \cdot atm}{K \cdot mol})} = 0.500 \text{ mole} \times \dfrac{2.02 \text{ g H}_2}{1 \text{ mol H}_2} = 1.01 \text{ g}$

5.31 $P_{total} = P_{water} + P_{oxygen} = 723 \text{ torr}$

$P_{water} = 23.8 \text{ torr}$

$P_{oxygen} = 723 - 23.8 = 699 \text{ torr}$

$699 \text{ torr} \times \dfrac{1 \text{ atm}}{760 \text{ torr}} = 0.920 \text{ atm}$

5.33   PV = nRT; Watch your units here!

$$n = \frac{PV}{RT} = \frac{(752 \text{ torr} \frac{1 \text{ atm}}{760 \text{ torr}})(6.00 \text{ L})}{(298 \text{ K})(0.0821 \frac{L \times atm}{K \times mol})} = 0.243 \text{ mol}$$

5.35 $n_{water} = 18.0 \text{ g} \times \dfrac{1 \text{ mol H}_2O}{18 \text{ g H}_2O} = 1 \text{ mol}$ $\qquad$ $n_{hydrogen} = 2.02 \text{ g} \times \dfrac{1 \text{ mol H}_2}{2.02 \text{ g H}_2} = 1 \text{ mol}$

$n_{carbon \; dioxide} = 44.0 \text{ g} \times \dfrac{1 \text{ mol CO}_2}{44.0 \text{ g CO}_2} = 1 \text{ mol}$

Temperature and pressure are the same for the three containers, therefore:

$$V_{H_2O} = V_{H_2} = V_{CO_2} = \frac{nRT}{P} = \frac{(1\,mol)(0.0821\frac{L \cdot atm}{K \cdot mol})(273\,K)}{1\,atm} = 22.4\,L$$

5.37 $P_{total} = P_{O_2} + P_{N_2} = 800\,torr$

$$P_{O_2} = x_{O_2}P_{total} \text{ where } x_{O_2} = \frac{n_{O_2}}{n_{O_2} + n_{N_2}}$$

$n_{O_2} = 0.250;\ n_{N_2} = 0.750$

Therefore, $P_{O_2} = 0.250(800\,torr) = 200\,torr.$

5.39 $V = \dfrac{nRT}{P} = \dfrac{(1.50\,mol)(0.0821\frac{L \cdot atm}{K \cdot mol})(328\,K)}{1.00\,atm} = 40.4\,L$

5.41 $\dfrac{P_1V_1}{T_1} = \dfrac{P_2V_2}{T_2} \rightarrow \dfrac{(565\,torr)(800\,mL)}{298\,K} = \dfrac{(760\,torr)(850\,mL)}{T_2}$

As long as the dimensions of the initial and final volumes and pressures are the same, this equation can be used directly.

$T_2 = 425\,K = 153°C.$

5.43 $\dfrac{P_1V_1}{T_1} = \dfrac{P_2V_2}{T_2} \rightarrow \dfrac{(600\,torr)(2.0\,L)}{298\,K} = \dfrac{(783\,torr)V_2}{398\,K}$

$V_2 = 2.0\,L$

5.45 $\dfrac{P_1V_1}{T_1} = \dfrac{P_2V_2}{T_2} \rightarrow \dfrac{(550\,torr)V_1}{323\,K} = \dfrac{(850\,torr)(1550\,mL)}{860\,K}$

$V_1 = 900\,mL$

5.47 $\text{density} = \dfrac{\text{mass}}{\text{volume}}; \quad \text{mass} = \text{\# moles} \times \text{molecular mass} \left[\dfrac{g}{mol}\right]$

$$\text{density} = \frac{n \times \text{molecular mass}}{\text{volume}} = \frac{n \times M}{V} = \frac{P \times M}{RT}$$

$$= \frac{(850 \text{ torr} \times \frac{1 \text{ atm}}{760 \text{ torr}})}{(303 \text{ K})(0.0821 \frac{\text{L} \cdot \text{atm}}{\text{K} \cdot \text{mol}})} \times \frac{2.02 \text{ g}}{\text{mol}} = 9.06 \times 10^{-2} \text{ g/L}$$

5.49 $Zn + 2HCl \rightarrow ZnCl_2 + H_2$

$$13.1 \text{ g Zn consumed} \times \frac{1 \text{ mol}}{65.4 \text{ g}} = 0.200 \text{ mol Zn} \times \frac{1 \text{ mol H}_2}{1 \text{ mol Zn}} = 0.200 \text{ mol H}_2$$

$$V = \frac{nRT}{P} = \frac{(0.200 \text{ mol})(0.0821\frac{\text{L} \cdot \text{atm}}{\text{K} \cdot \text{mol}})(273 \text{ K})}{1 \text{ atm}} = 4.48 \text{ L}$$

$$4.48 \text{ L} \times \frac{1000 \text{ mL}}{\text{L}} = 4.48 \times 10^3 \text{ mL}$$

5.51   (a) $0.750 \text{ atm} \times \frac{760 \text{ torr}}{1 \text{ atm}} = 570 \text{ torr.}$

(b) $1.23 \text{ atm} \times \frac{760 \text{ torr}}{1 \text{ atm}} = 935 \text{ torr.}$

(c) $0.950 \text{ atm} \times \frac{760 \text{ torr}}{1 \text{ atm}} = 722 \text{ torr.}$

(d) $1.750 \text{ atm} \times \frac{760 \text{ torr}}{1 \text{ atm}} = 1330 \text{ torr.}$

5.53   Use $PV = nRT$ to get:

(a) 27.7 L   (b) 23.1 atm   (c) 329 K   (d) 1.00 mole

5.55 $\dfrac{P_1}{n_1} = \dfrac{P_2}{n_2}$

$$n_1 = \frac{P_1 V_1}{RT_1} = \frac{(14.0 \text{ atm})(20.0 \text{ L})}{(300 \text{ K})(0.0821 \frac{\text{L} \cdot \text{atm}}{\text{K} \cdot \text{mol}})} = 11.37 \text{ mol}$$

$$n_2 = \left(\frac{11.5 \text{ atm}}{14.0 \text{ atm}}\right)(11.36 \text{ mol}) = 9.33 \text{ mol}$$

Number of moles of O_2 removed = $n_1 - n_2 = 2.04 \text{ mol}$

5.57 $C_H = \dfrac{0.51 \text{ mL CO}_2}{\text{mL blood}}$ at 1.0 atm.

$$\frac{V_{CO_2}}{V_{blood}} = C_H P_{CO_2} \rightarrow V_{CO_2} = \left(\frac{0.51 \text{ mL CO}_2}{\text{mL blood}}\right)(1 \text{ mL blood})\left(\frac{46 \text{ torr}}{760 \text{ torr}}\right)$$

$= 0.031 \text{ mL CO}_2$

5.59 $P_1V_1 = P_2V_2$

(4.25 atm) (0.0850 mL) = (1.00 atm) (V_2) $\rightarrow$ $V_2 = 0.0361$ mL

5.61 $n_1 = \dfrac{P_1V_1}{RT_1} = \dfrac{(7.500 \text{ atm})(50.00 \text{ L})}{(298.0 \text{ K})(0.0821 \frac{L \cdot atm}{K \cdot mol})} = 15.32 \text{ mol} \times \dfrac{44.09 \text{ g}}{\text{mol}} = 675.8 \text{ g}$

5.63 $n_1 = \dfrac{P_1V_1}{RT_1} = \dfrac{(4.000 \times 10^{-7} \text{ atm})(0.001000 \text{ L})}{(250 \text{ K})(0.0821 \frac{L \cdot atm}{K \cdot mol})} = 1.95 \times 10^{-11} \text{ mol} \times \dfrac{6.022 \times 10^{23} \text{ molecule}}{\text{mol}} =$

1.17×10^{13} molecules in 1.0 mL

5.65

$$\frac{n_1}{V_1} = \frac{P_1}{RT_1} = \frac{(1 \text{ atm})}{(373 \text{ K})(0.0821 \frac{L \cdot atm}{K \cdot mol})} = 3.27 \times 10^{-2} \frac{mol}{L} \times \frac{18.02 \text{ g}}{mol} \times \frac{1 L}{1000 \text{ mL}} \rightarrow d = 5.89 \times 10^{-4} \text{ g/mL}$$

5.67   62.6 mg of that material is equal to

$$n_1 = \frac{P_1V_1}{RT_1} = \frac{(772 \text{ torr})(0.0349 \text{ L})}{(373 \text{ K})(0.0821 \frac{L \cdot atm}{K \cdot mol})} \times \frac{1 \text{ atm}}{760 \text{ torr}} = 1.16 \times 10^{-3} \text{ mol}$$

$$\frac{0.0626 \text{ g}}{1.16 \times 10^{-3} \text{ mol}} = 54.0 \text{ g/mol}$$

0.8882 × 54.0 = 48.0 g/mol is the carbon content of the material, which equals 4 carbon atoms. 0.1118 × 54 = 6 g/mol is the hydrogen content of the material, which equals 6 hydrogen atoms. Thus, the molecular formula is C_4H_6.

5.69 $n_1 = \dfrac{P_1V_1}{RT_1} = \dfrac{(8.00 \text{ atm})(65.0 \text{ L})}{(298 \text{ K})(0.0821 \frac{L \cdot atm}{K \cdot mol})} = 21.25 \text{ mol} \times \dfrac{44.0 \text{ g}}{\text{mol}} = 935.1 \text{ g}$

$n_2 = \dfrac{P_2V_2}{RT_2} = \dfrac{(5.10 \text{ atm})(65.0 \text{ L})}{(298 \text{ K})(0.0821 \frac{L \cdot atm}{K \cdot mol})} = 13.55 \text{ mol} \times \dfrac{44.0 \text{ g}}{\text{mol}} = 596 \text{ g}$

Thus, 3.39×10^2 g of CO_2 was used.

5.71 $$\frac{n_1}{V_1} = \frac{P_1}{RT_1} = \frac{(1\,atm[=760\,torr])}{(273\,K)(0.0821\,\frac{L \cdot atm}{K \cdot mol})} = 4.46 \times 10^{-2}\,\frac{mol}{L}$$

$$\frac{1.071\,g}{1\,L} \times \frac{1\,L}{4.46 \times 10^{-2}\,mol} = 24.0\,g/mol$$

Thus, the average molecular weight of the gas is 24 g/mol. By setting up and solving the following relations:

$32x + 16(1-x) = 24 \rightarrow x = 0.5$

Thus, the fraction of moles of oxygen is equal to the fraction of moles of methane and the ratio of the oxygen and methane is $O_2/CH_4 = 1/1$.

5.73   Every dive adds nitrogen to the blood. Making the first of a series of dives the deepest, maximizes the calculation of blood nitrogen accumulation. Making it the last dive fools the diver into a false sense of security and could possibly bring on a serious case of the bends upon surfacing.

Chapter 6

Interactions Between Molecules

6.1   All matter would exist only in the gaseous state. No matter would be present in either liquid or solid form.

6.3   The melting of a solid describes the transition from a structure consisting of a highly ordered array of molecules to a disordered structure with the molecules still in contact but now free to move about. This transition is effected by adding energy to the system to overcome the strong attractive forces in the solid.

6.5   True. The vapor pressure of a liquid is a measure of the escaping tendency of its molecules. Increasing the temperature increases the kinetic energy of its molecules thus increasing their escaping tendency.

6.7   Because the molecules are in contact, so there is no possibility of a change in volume with an increase in pressure.

6.9   The molar heat of vaporization is always greater than the molar heat of fusion, because to melt a solid only a portion of the attractive forces must be overcome, while to vaporize a liquid, all secondary attractive forces must be overcome.

6.11   True. The transformation between the solid state and liquid state occurs at the same temperature. If it is approached by raising the temperature of the solid until the liquid forms, it is called the melting point. If it is approached by lowering the temperature of the liquid until the solid forms, it is called the freezing point.

6.13   9.7 kcal of heat will be evolved because the condensation of a gas is merely the reverse of vaporization.

6.15   Through the weak London forces (temporary dipoles).

6.17   Hydrogen bonds require an H covalently bonded to an O, N, or F. (b) is the only pair where this occurs in both compounds.

6.19   The greater the number of electrons (atomic number) in a molecule, the larger the temporary dipole (London force). This means that in general, the greater the molecular mass the greater the strength of London forces exerted between molecules.

6.21.   The four C-Cl polar bonds are arranged symmetrically (tetrahedrally) with carbon at the center of the tetrahedron. The symmetric arrangement of dipoles results in a net dipole moment of zero.

6.23

	London force	Dipole-dipole	Hydrogen bond
CH_4	Yes	No	No
$CHCl_3$	Yes	Yes	No
NH_3	Yes	Yes	Yes

6.25   (a) has the greater heat of vaporization because it can form hydrogen bonds between molecules, whereas (b) cannot.

6.27   The molecules of a liquid have acquired sufficient kinetic energy to partially overcome the secondary attractive forces holding them in fixed positions. While they are free to assume new positions under an applied force (e. g., gravity), they still remain in contact.

6.29   In an open system vapor molecules will continuously escape into the surroundings. Therefore the rate of evaporation will always exceed the rate of condensation, and evaporation will occur continuously.

6.31   When a solid melts, the ordered solid state structure becomes disordered, but the molecules in the liquid state remain in close contact. Therefore the volume of the liquid is close to that of the solid.

6.33   The word "vapor" is used to describe the gaseous state of a substance whose liquid and gaseous states are present at the same time. Under normal conditions of room temperature, the liquid states of oxygen or nitrogen are not possible.

6.35   b > a > c... these are ordered by increasing strength of secondary forces, ionic > hydrogen bond > dipole-dipole forces.

6.37   No. Vapor pressure is a balance between evaporation and condensation. In an open container the rate of evaporation is always greater than the rate of condensation, and equilibrium cannot be established.

6.39   Equilibrium describes a situation that does not change over time.

6.41   The temperature at which vapor bubbles form throughout a liquid at an external pressure of one atmosphere.

6.43   Both molecules are polar, but additionally, water can form hydrogen bonds, so the secondary attractive forces in water are greater than those in chloroform. Therefore water will have the greater surface tension of the two.

6.45   Acetic acid molecules form hydrogen bonds among themselves and also to water. Pentane molecules can only interact by London forces. The attractive forces between acetic acid and water are similar and therefore acetic acid will dissolve in water. The attractive forces between pentane molecules and water are very different and will therefore not form solutions.

6.47   An amorphous solid has no particular shape, melts over a range of temperatures, and if it can be shattered, forms pieces that have round or smooth surfaces resembling a liquid.

6.49 $CH_4 < C_2H_5OH < NaCl$. These are ordered by decreasing strength of secondary attractive forces, London dispersion < hydrogen bond < ionic forces.

6.51 $\dfrac{21.6}{31.8} = .68 = 68\%$

6.53   A molecule that has polar bonds and is not symmetric will exhibit a dipole. Examples include H_2O, HF, and $CHCl_3$.

6.55 H_2O can form two hydrogen bonds per molecule, while HF can form only one hydrogen bond per molecule. Therefore, more thermal energy is required to break two hydrogen bonds than one. This results in a higher boiling point.

6.57   Vapor pressure measures the tendency of molecules to escape from the liquid into the gas phase. The greater the secondary forces, the lower the escaping tendency and the smaller the vapor pressure.

6.59   The Lewis Dot structure of dimethyl ether shows that the oxygen has a free pair of electrons. This free pair of electrons acts to force the molecule into a bent configuration like water. Thus, dimethyl ether has a dipole.

6.61   Both water and hydrogen sulfide can form dipole-dipole and London secondary attractive forces, but water can form hydrogen bonds, hydrogen sulfide cannot. The strong hydrogen bonds lead to water being a solid at the same temperature that hydrogen sulfide exists as a gas.

6.63   The temperature on the outside of an uninsulated glass of ice and water is about 0°C, the freezing point of water. Water vapor in the air will therefore condense on the outside of the glass.

6.65   (a) HF forms hydrogen bonds, HCl cannot.

        (b) $TiCl_4$ condenses by London forces; the liquid form of LiCl consists of ions, which condense by ionic forces.

        (c) HCl condenses through both dipole-dipole and London forces, the sum of which is weaker than the ionic forces in liquid LiCl.

        (d) The hydrogen bonds formed in ethanol are stronger than the secondary forces in ethyl ether.

6.67   When a substance melts, the secondary forces are only loosened so that the crystalline order is disrupted. Vaporization requires all secondary forces to be broken, and this requires more energy.

6.69   Both a liquid and a gas are disordered compared to a solid.

6.71   The condensation of steam (gaseous water) releases far more heat than the cooling of liquid water from 100°C to skin temperature.

6.73   Without lung surfactant, the surface tension at the surface of alveoli would cause the alveoli to collapse and thus prevent absorption of oxygen and diffusion of carbon dioxide.

Chapter 7

Solutions

7.1 (a) $25.2 \text{ g solution} \times \dfrac{4.25 \text{ g solute}}{100 \text{ g solution}} = 1.07 \text{ g solute}$

(b) $125 \text{ g solution} \times \dfrac{6.055 \text{ g solute}}{100 \text{ g solution}} = 8.19 \text{ g solute}$

7.3 $0.600 \text{ w/v\% solution means } \dfrac{\text{grams of solute}}{\text{mL of solution}} = \dfrac{0.600}{100} = .00600$

$\dfrac{x \text{ g KHCO}_3}{250 \text{ mL solution}} = 0.00600 \rightarrow x = 1.50 \text{ g KHCO}_3$

7.5 $6.20 \text{ w/v\% solution means } \dfrac{\text{grams of solute}}{\text{mL of solution}} = \dfrac{6.20}{100} = 0.0620$

$\dfrac{5.40 \text{ g NH}_4\text{Cl}}{x \text{ mL solution}} = 0.0620 \rightarrow x = 87.1 \text{ mL of solution}$

7.7 $5.00 \text{ w/w\% solution means } \dfrac{\text{grams of solute}}{\text{grams of solution}} = \dfrac{5.00}{100} = .0500$

$\dfrac{6.25 \text{ g glucose}}{x \text{ g solution}} = 0.0500 \rightarrow x = 125 \text{ g of solution.}$

7.9 $7.50 \text{ v/v\% solution means } \dfrac{\text{volume of solute}}{\text{volume of solution}} = \dfrac{7.50}{100} = 0.07500$

$\dfrac{33.6 \text{ mL proylene glycol}}{x \text{ mL solution}} = 0.0750 \rightarrow x = 448 \text{ mL of solution.}$

7.11 $\text{mg\% solution means } \dfrac{\text{mg of solute}}{\text{mL of solution}} \times 100$

$\dfrac{0.230 \text{ mg}}{19.0 \text{ mL}} \times 100 = 1.21 \text{ mg\%}$

7.13 $42 \text{ ppm} = \dfrac{42 \text{ mg of solute}}{1 \text{ L of solution}}$ $\% \text{ w/v} = \dfrac{\text{g of solute}}{100 \text{ mL of solution}}$

$42 \text{ mg} \times \dfrac{1 \text{ g}}{1000 \text{ mg}} = .042 \text{ g}$ $1 \text{ L} \times \dfrac{1000 \text{ mL}}{1 \text{ L}} = 1000 \text{ mL}$

$\dfrac{0.042 \text{ g}}{1000 \text{ mL}} \times 100 = .0042 \% \text{ w/v}$

7.15

$2.5 \text{ L of a } 0.40 \text{ M solution} = 2.5 \text{ L} \times \dfrac{0.40 \text{ mol of solute}}{1 \text{ L of solution}} = 1 \text{ mol of solute}$

7.17 $\text{Molarity} = [\text{M}] = \dfrac{\text{moles of solute}}{\text{L of solution}}$

(a) $\dfrac{2.6 \text{ mol of solute}}{1.3 \text{ L of solution}} = 2.0 \text{ M}$

(b) $\dfrac{0.810 \text{ mol of solute}}{0.240 \text{ L of solution}} = 3.38 \text{ M}$

7.19 $\text{Moles of solute} = (\text{L of solution}) \times (\text{molarity})$

(a) $(1.05 \text{ L}) \times (2.65 \text{ M}) = 2.78 \text{ mol}$

(b) $(0.452 \text{ L}) \times (0.850 \text{ M}) = 0.384 \text{ mol}$

7.21 $\text{Molarity} = [\text{M}] = \dfrac{\text{moles of solute}}{\text{L of solution}}$; therefore volume of solution $= \dfrac{\text{moles of solute}}{\text{molarity}}$

(a) $\dfrac{1.35 \text{ mol of solute}}{0.90 \text{ M}} = 1.50 \text{ L of solution}$

(b) $\dfrac{2.80 \text{ mol of solute}}{0.731 \text{ M}} = 3.83 \text{ L of solution}$

7.23 $23.83 \text{ mg MgCl}_2 \times \dfrac{1 \text{ g}}{1000 \text{ mg}} \times \dfrac{1 \text{ mol}}{95.2 \text{ g}} = 0.000250 \text{ mol MgCl}_2$

$\dfrac{0.00025 \text{ mol of solute}}{0.275 \text{ L of solution}} = 9.09 \times 10^{-4} \text{ M}$

7.25 $0.450 \text{ M} \times 0.0750 \text{ L} = 0.0338 \text{ mol CaCl}_2 \times \dfrac{111.0 \text{ g}}{\text{mol}} \times \dfrac{1000 \text{ mg}}{\text{g}} = 3.75 \times 10^3 \text{ mg}.$

7.27 $25.0 \text{ g KCl} \times \dfrac{1 \text{ mol}}{74.6 \text{ g}} = 0.335 \text{ mol}$

$\dfrac{0.335 \text{ mol KCl}}{0.620 \dfrac{\text{mol KCl}}{\text{L}}} = 0.540 \text{ L} \times \dfrac{1000 \text{ mL}}{\text{L}} = 5.40 \times 10^2 \text{ mL}$

7.29 $V_1 M_1 = V_2 M_2$

$V_1 = ?$ $M_1 = 36 \text{ M}$

$V_2 = 6.0 \text{ L}$ $M_2 = 0.12 \text{ M}$

$V_1 = \dfrac{6.0 \text{ L} \times 0.12 \text{ M}}{36 \text{ M}} = 0.020 \text{ L}$

7.31 $V_1 M_1 = V_2 M_2$

(a) $V_1 = ?$ $M_1 = 0.750 \text{ M}$

$V_2 = 4.50 \text{ L}$ $M_2 = 0.250 \text{ M}$

$V_1 = \dfrac{4.50 \text{ L} \times 0.250 \text{ M}}{0.75 \text{ M}} = 1.50 \text{ L}$

(b) $V_1 = ?$ $M_1 = 0.360 \text{ M}$

$V_2 = 650 \text{ mL}$ $M_2 = 0.180 \text{ M}$

$V_1 = \dfrac{650 \text{ mL} \times 0.180 \text{ M}}{0.360 \text{ M}} = 325 \text{ mL}$

7.33 $V_1 M_1 = V_2 M_2$

$V_1 = 0.80 \text{ L}$ $M_1 = ?$

$V_2 = 4.0 \text{ L}$ $M_2 = 1.8\% \text{ w/v}$

$M_1 = \dfrac{4.0 \text{ L} \times 1.8 \% \text{ w/v}}{0.80 \text{ L}} = 9.0\% \text{ w/v}$

7.35 $V_1M_1 = V_2M_2$

$V_1 = 0.100$ L $M_1 = 2.10$ M

$V_2 = 0.420$ L $M_2 = $?

$$M_2 = \frac{0.100 \text{ L} \times 2.10 \text{ M}}{0.420 \text{ L}} = 0.500 \text{ M}$$

7.37   No. If the solution at 10°C were saturated it would be at a concentration on 43 g/100 mL. When the temperature of that solution is raised to 50°C, the largest amount of strontium acetate that could stay in solution is 37.4 g/100 mL. Therefore, to attain this concentration, 5.6 g/100 mL of strontium acetate would have to precipitate out of solution. Since the solution remained clear, it must not have been saturated at 10°C.

7.39 $\text{Molarity} = [M] = \dfrac{\text{moles of solute}}{\text{Liter of solution}}$

$74.5 \text{ g} \times \dfrac{1 \text{ mol}}{111.1 \text{ g}} = .671 \text{ mol}$ $100 \text{ mL} \times \dfrac{1 \text{ L}}{1000 \text{ mL}} = 0.100 \text{ L}$

$\dfrac{0.671 \text{ mol}}{0.100 \text{ L}} = 6.71 \text{ M}$

7.41   (a) $AlCl_3$ ionizes to Al^{3+} + 3 Cl^-, therefore $i = 4$; 0.10 M × 4 = 0.40 osmol/L

(b) KCl ionizes to K^+ + Cl^-, therefore $i = 2$; 0.01 M × 2 = 0.02 osmol/L

7.43   From Box 7.3 a solution must be 0.3 osmol/L to be isotonic with red blood cells

$8.0\% \text{ w/v glucose solution} = \dfrac{8.0 \text{ g}}{100 \text{ mL}} \text{ NaCl solution}$

Glucose does not dissociate in solution.

$\dfrac{8.0 \text{ g glucose}}{100 \text{ mL}} \times \dfrac{1000 \text{ mL}}{\text{L}} \times \dfrac{1 \text{ mol}}{180 \text{ g}} = 0.44 \dfrac{\text{mol}}{\text{L}} \times \dfrac{1 \text{ ion}}{\text{mol}} = 0.44 \dfrac{\text{osmol}}{\text{L}}$

0.44 osmol/L > 0.3 osmol/L. Therefore water will flow from the cell into the solution, the erythrocytes will lose water and can shrink.

7.45   (a) $\dfrac{13.0 \text{ g CaCl}_2}{13.0 \text{ g} + 133 \text{ g H}_2\text{O}} = 0.0890 \times 100 = 8.90\% \text{ w/w}$

(b) $\dfrac{12.5 \text{ g ethyl alcohol}}{12.5 \text{ g} + 165 \text{ g H}_2\text{O}} = 0.0704 \times 100 = 7.04\% \text{ w/w}$

(c) $\dfrac{2.5 \text{ g}}{2.5 \text{ g} + 60 \text{ g}} = 0.040 \times 100 = 4.0\% \text{ w/w}$

7.47 (a) 55.0 g of solution $\times \dfrac{4.00 \text{ g glucose}}{100 \text{ g of solution}} = 2.20$ g of glucose.

(b) 175 g of solution $\times \dfrac{7.50 \text{ g ammonium sulfate}}{100 \text{ g of solution}} = 13.1$ g of ammonium sulfate.

7.49 (a) $\dfrac{6.40 \text{ g}}{66.0 \text{ mL}} \times 100 = 9.70\% \text{ w/v}$ (b) $\dfrac{1.30 \text{ g}}{125 \text{ mL}} \times 100 = 1.04\% \text{ w/v}$

(c) $\dfrac{3.60 \text{ g}}{90.0 \text{ mL}} \times 100 = 4.00\% \text{ w/v}$

7.51 Molarity = [M] = $\dfrac{\text{moles of solute}}{\text{L of solution}}$

Therefore # of moles in solution = (molarity) $\times$ (L of solution).

(a) 0.66 M $\times$ .046 L = 0.0030 mol   (b) 0.065 M $\times$ 1.2 L = 0.078 mol

(c) 0.250 M $\times$ 0.495 L = 0.124 mol   (d) 1.75 M $\times$ 0.625 L = 1.09 mol

7.53 0.375 M = $\dfrac{0.375 \text{ mol of glucose}}{\text{L of solution}}$

(a) $\dfrac{0.375 \text{ mol of glucose}}{\text{L of solution}} \times 1.49 \text{ L} = 0.559$ mol

(b) $\dfrac{0.375 \text{ mol of glucose}}{\text{L of solution}} \times 6.72 \text{ L} = 2.52$ mol

(c) $\dfrac{0.375 \text{ mol of glucose}}{\text{L of solution}} \times 2.56 \text{ L} = 0.960$ mol

(d) $\dfrac{0.375 \text{ mol of glucose}}{\text{L of solution}} \times 1.81 \text{ L} = 0.679$ mol

7.55 0.160 M = $\dfrac{0.160 \text{ mol of KI}}{\text{L of solution}} \times 0.500 \text{ L} = 0.0800$ mol KI needed

0.0800 mol KI $\times \dfrac{166 \text{ g}}{1 \text{ mol}} = 13.3$ g of KI needed.

Therefore, add water to 13.3 g of KI to a final volume of 500.0 mL.

7.57   (a) $\dfrac{15.0 \text{ g CaCl}_2}{15.0 \text{ g} + 250 \text{ g H}_2\text{O}} = 0.0566 \times 100 = 5.66\%$ w/w

(b) $\dfrac{16.5 \text{ g NaCl}}{16.5 \text{ g} + 325 \text{ g H}_2\text{O}} = 0.0483 \times 100 = 4.83\%$ w/w

(c) $\dfrac{2.20 \text{ g C}_2\text{H}_5\text{OH}}{2.20 \text{ g} + 122 \text{ g H}_2\text{O}} = 0.0177 \times 100 = 1.77\%$ w/w

7.59   (a) $0.950 \text{ L} \times \dfrac{1.25 \text{ mol NaCl}}{\text{L}} \times \dfrac{58.44 \text{ g}}{1 \text{ mol}} = 69.5$ g NaCl

(b) $625 \text{ mL} \times \dfrac{1 \text{ L}}{1000 \text{ mL}} \times \dfrac{0.750 \text{ mol Mg(NO}_3)_2}{\text{L}} \times \dfrac{148.3 \text{ g}}{1 \text{ mol}} = 69.5$ g

(c) $325 \text{ mL} \times \dfrac{1 \text{ L}}{1000 \text{ mL}} \times \dfrac{2.20 \text{ mol K}_2\text{CO}_3}{\text{L}} \times \dfrac{138.2 \text{ g}}{1 \text{ mol}} = 98.7$ g

7.61 $V_1 M_1 = V_2 M_2$

$V_1 = 1.50$ L $M_1 = 1.80$ M

$V_2 = ?$ $M_2 = 0.180$

$V_2 = \dfrac{1.50 \text{ L} \times 1.80 \text{ M}}{0.180 \text{ M}} = 15.0$ L

7.63 $V_1 M_1 = V_2 M_2$

$V_1 = 0.150$ L $M_1 = ?$

$V_2 = 1.50$ L $M_2 = 0.90\%$ w/v

$M_1 = \dfrac{1.50 \text{ L} \times 0.90 \% \text{ w/v}}{0.150 \text{ L}} = 9.0\%$ w/v

7.65   From Table 7.3:

(a) AgCl is not soluble in H_2O, therefore a precipitate forms:

$$Ag^+(aq) + Cl^-(aq) \rightarrow AgCl(s)$$

(b) $Mg(OH)_2$ is not soluble in H_2O, therefore a precipitate forms:

$$Mg^{2+}(aq) + 2\ OH^-(aq) \rightarrow Mg(OH)_2(s)$$

(c) $PbCl_2$ is not soluble in H_2O, therefore a precipitate forms:

$$Pb^{2+}(aq) + 2Cl^-(aq) \rightarrow PbCl_2(s)$$

(d) $Ca_3(PO_4)_2$ is not soluble in H_2O, therefore a precipitate forms:

$$3\ Ca^{2+}(aq) + PO_4^{3-}(aq) \rightarrow Ca_3(PO_4)_2\ (s)$$

7.67   In a solution, the minor component is the solute and the major component is the solvent. Therefore, propylene glycol is the solvent and water is the solute.

7.69   100 torr > 15 torr, therefore water moves out of the arterial blood.

7.71   From Box 7.3, 0.30 Osmol/L.

7.73   Assume 100 cm^3 of wine, which is 12.5% ethanol by volume. Thus, this 100 cm^3 will consist of 87.5 cm^3 of water and 12.5 cm^3 of ethanol.

$$12.5\ cm^3 \times \frac{0.79\,g}{1\ cm^3} = 9.87\,g \text{ ethanol and } 87.5\ cm^3 \times \frac{1.0\,g}{1\ cm^3} = 87.5\,g \text{ water}$$

$$\frac{9.87\,g}{9.87 + 87.5\,g} = 10.1\%\ w/w$$

7.75   The net osmotic pressure is 0.100 M, due to the difference in the concentrations.

$\Pi = nRT/V = 0.100 \times R \times T = 2.45$ atm
1 atm = 760 mm Hg, 2.45 atm = 1862 mm Hg
Hg density = 13.6 times H_2O density
1862 mm Hg = 13.6 × 1862 mm H_2O = 25,323 mm H_2O = 25.3 m H_2O

Thus, the water will rise 25.3 meters.

7.77   To determine the molarity of this solution, you need to calculate the number of moles of sulfuric acid in 1 liter. Thus, assume that you have 1.00 L of this solution, which is 98.0% by mass H_2SO_4.

$$1.00\ L \times \frac{1.84\,g}{1\ cm^3} \times \frac{1000\,cm^3}{1\ L} = 1840\,g\ H_2SO_4 \times \frac{1\,mol}{98\ g\ H_2SO_4} = 18.4\ mol$$

18.4 mol in one liter, which equals 18.4 M

7.79   (a) $Na_2SO_4(aq) + BaCl_2(aq) \rightarrow 2\ NaCl(aq) + BaSO_4(s)$

$$Ba^{2+}(aq) + SO_4^{2-}(aq) \rightarrow BaSO_4(s)$$

(b) $NaCl(aq) + AgNO_3(aq) \rightarrow NaNO_3(aq) + AgCl(s)$

$$Ag^+(aq) + Cl^-(aq) \rightarrow AgCl(s)$$

7.81   At solution concentrations greater than the critical micelle concentration, surface-active substances associate through secondary forces to form large-scale structures called micelles.

Chapter 8

Chemical Reactions

8.1 $3 O_2 (g) \rightarrow 2 O_3 (g)$

The ratio of O_3 to O_2 is $\dfrac{2 \text{ mol } O_3}{3 \text{ mol } O_2}$

$\dfrac{1.50 \text{ mol } O_3}{1 \text{ min}} \times \dfrac{3 \text{ mol } O_2}{2 \text{ mol } O_3} = \dfrac{2.25 \text{ mol } O_2 \text{ (g)}}{\text{min}}$

8.3   The increase in temperature results in a proportional increase in the number of molecules that experience collisions that have enough energy to result in reaction.

8.5   For reaction (a) to occur, the two reactants must have kinetic energies greater than 24 kJ, whereas for reaction (b) to occur the two reactants must have kinetic energies greater than 53 kJ. At a given temperature, the percentage of molecules that have K.E. > 24 kJ will be greater than the percentage that have K.E. > 53 kJ. Therefore, the rate of reaction (a) will be greater than the rate of reaction (b).

8.7 $\Delta H_{reaction} = E_{forward} - E_{back}$

$-21 \text{ kJ} = 37 \text{ kJ} - E_{back}$

$E_{back} = 58 \text{ kJ}$

8.9   No. The reaction takes place in an open system so that the CO_2 escapes to the atmosphere. Therefore, the CO_2 that is formed cannot react with CaO to allow the reverse reaction to occur. Therefore, the system can never come to equilibrium.

8.11   Forward reaction: $CaCO_3(s) \rightarrow CaO(s) + CO_2(g)$

Reverse reaction: $CaO(s) + CO_2(g) \rightarrow CaCO_3(s)$

8.13   Pure solids and liquids do not appear in equilibrium constants, therefore, $K_{eq} = [CO_2]$.

8.15 $K_{eq} = \dfrac{[HI]^2}{[H][I]} = \dfrac{(0.27)^2}{(0.86)(0.86)} = .099$

8.17   LeChatelier's Principle states that when a reactant or product is removed from a reaction at equilibrium, the equilibrium shifts to replace the compound that was removed.

$$N_2(g) + 3H_2(g) \leftrightarrow 2NH_3$$

If NH_3 is removed, the reaction shifts to replace NH_3, which means that it shifts to the products.

8.19   (a) LeChatelier's Principle states that when the temperature is lowered in a reaction at equilibrium, the equilibrium shifts to raise the temperature. In this reaction that means that the reaction shifts to the reactants which will result in a color shift to pink.

(b) LeChatelier's Principle states that when a reactant or product is added to a reaction at equilibrium, the equilibrium shifts to use up the compound that was added. In this reaction when Cl^- is added, the reaction shifts to use up more Cl^-; the reaction will shift to the products which will result in a color shift to blue.

8.21   The square brackets around the chemical compounds in an equilibrium constant expression denote their molar concentrations.

8.23   The concentrations of pure liquids and solids do not appear in K_{eq}, therefore:

(a) $K_{eq} = \dfrac{[CO][H_2]^3}{[CH_4][H_2O]}$ 　　　　(b) $K_{eq} = \dfrac{[CO_2]^8[H_2O]^8}{[O_2]^{12}}$

8.25   The concentrations of pure liquids and solids do not appear in K_{eq}, therefore:

(a) $K_{eq} = [CO_2][NH_3]^2$ 　　　　(b) $K_{eq} = \dfrac{1}{[O_2]^{0.5}}$

8.27   (a) $COCl_2(g) \leftrightarrow CO(g) + Cl_2(g)$

(b) $CH_4(g) + H_2O(g) \leftrightarrow CO(g) + 3\ H_2(g)$

8.29   The concentrations of pure liquids and solids do not appear in K_{eq}, therefore:

(a) $K_{eq} = \dfrac{[NO_2][NO_3]}{[N_2O_5]}$ 　　　　(b) $K_{eq} = \dfrac{[N_2]^2[H_2O]^6}{[NH_3]^4[O_2]^3}$

8.31 $A(g) \leftrightarrow B(g)$

$K_{eq} = \dfrac{[B]}{[A]}$; Therefore at equilibrium $\dfrac{[A]}{[B]} = \dfrac{1}{K_{eq}}$

(a) 1000/1 　　　　(b) 1/100 　　　　(c) 7.1/1 　　　　(d) 12/1

8.33   The dissolution of NH_4CNS can be thought of as the reaction

$$NH_4CNS(s) \leftrightarrow NH_4CNS(aq)$$

The solution becomes cold when this "reaction" goes to the right, therefore heat is needed to allow this reaction to occur.

$$NH_4CNS(s) + heat \leftrightarrow NH_4CNS(aq)$$

If the temperature is raised LeChatelier's principle tells us that the reaction will shift to lower the temperature. In this reaction, that means that the reaction shifts to the right and more solid will dissolve.

8.35 $\Delta H_{reaction} = E_{forward} - E_{back}$

$15 \text{ kJ} = E_{forward} - 57 \text{ kJ}$

$E_{forward} = 72 \text{ kJ}$

8.37 $\Delta H_{reaction} = E_{forward} - E_{back}$

since $\Delta H > 0$, $E_{forward} > E_{back}$

8.39   If a system in an equilibrium state is disturbed, the system will adjust to neutralize that disturbance and restore the system to equilibrium.

8.41   As the reaction proceeds, the total number of moles decreases. Since all components are gaseous, follow the course of the reaction by measuring the decrease in pressure as a function of time.

8.43   From Figure 8.6, decreasing the temperature 20°C will approximately halve the reaction rate. 98.6°F − 72°F = 26.6°F ≈ 15°C. Thus, the rate is decreased by approximately a factor of 1.5, which means the pulse rate at 72°F is 40 ± 5 beats per minute.

8.45   The easiest approach would be to carry out the reaction in the dark. Another approach would be to use vessels made with intensely tinted glass. A more sophisticated method would be to use vessels made with intensely tinted glass of different colors.

8.47   (a) LeChatelier's Principle states that when a reactant or product is added to a reaction at equilibrium, the equilibrium shifts to remove the compound that was added, thus this reaction shifts to the left (reactants).

(b) LeChatelier's Principle states that when the pressure is increased in a reaction at equilibrium, the reaction will shift to decrease the pressure, which for this reaction is to shift to the side of the reaction with fewer gases, which is to the left (reactants).

(c) LeChatelier's Principle states that when the pressure is increased in a reaction at equilibrium, the reaction will shift to decrease the pressure, which for this reaction is to shift to the side of the reaction with fewer gases, which is to the left (reactants).

(d) LeChatelier's Principle states that when the temperature is lowered in a reaction at equilibrium, the equilibrium shifts to raise the temperature. In this reaction, that means that the reaction shifts to the left (reactants).

(e) No effect.

(f) LeChatelier's Principle states that when a reactant or product is removed from a reaction at equilibrium, the equilibrium shifts to replace the compound that was removed, thus this reaction will shift to the right (products).

8.49   An overall equilibrium constant for a sequence of reactions can be calculated only if a product of the first reaction is a reactant of the next reaction, and so on for each succeeding reaction.

8.51 E_a of the uncatalyzed reaction is larger. Catalysts provide an alternate path for a reaction of lower E_a, which results in faster rate. ΔH is unchanged because the reactants and products are the same.

8.53 $K = \dfrac{[CO(g)][Cl_2(g)]}{[CoCl_2(g)]} = \dfrac{0.046M \times 0.046M}{0.500M - 0.046M} = 0.0047$

Chapter 9

Acids, Bases, and Buffers

9.1 The ion product, $K_w = [H_3O^+] \times [OH^-] = (1.00 \times 10^{-7}) \times (1.00 \times 10^{-7}) = 1.00 \times 10^{-14}$

9.3 An amphoteric substance is a compound that can act as an acid or a base.

9.5 A strong base is dissociated 100% in aqueous solution. Three examples are NaOH, KOH and LiOH.

9.7 0.30 M HCl solution. HCl is a strong acid, which means that every HCl molecule dissociates into an H_3O^+ ion and a Cl^- ion. Thus, $[H_3O^+] = [Cl^-] = 0.30$ M.

9.9 0.25 M TlOH solution. TlOH is a strong base, which means that every TlOH molecule dissociates into an Tl^+ ion and an OH^- ion. Thus, $[Tl^+] = [OH^-] = 0.25$ M.

9.11 0.350 M HCl solution. HCl is a strong acid, which means that every HCl molecule dissociates into an H_3O^+ ion and a Cl^- ion. Thus, $[H_3O^+] = [Cl^-] = 0.350$ M. It is also known that

$$[H_3O^+] \times [OH^-] = 1 \times 10^{-14} \rightarrow [OH^-] = \frac{1 \times 10^{-14}}{0.350} = 2.86 \times 10^{-14}$$

9.13 0.020 M $Ca(OH)_2$ solution. $Ca(OH)_2$ is a strong base, which means that every $Ca(OH)_2$ molecule dissociates into an Ca^{2+} ion and 2 OH^- ions. Thus, $[Ca^{2+}] = 0.020$ M. Because there are twice as many OH^- ions in solution than Ca^{2+} ions, its concentration must be twice that of Ca^{2+}, Thus, $[OH^-] = 0.020$ M $\times 2 = 0.040$ M. It is also known that

$$[H_3O^+] \times [OH^-] = 1 \times 10^{-14} \rightarrow [H_3O^+] = \frac{1 \times 10^{-14}}{0.040} = 2.5 \times 10^{-13}$$

9.15 The pH of pure water at 25°C is 7.00.

9.17 $pH = -\log [H_3O^+] = -\log (0.010) = 2.00$

9.19 $pOH = -\log[OH^-] = -\log(0.015) = 1.82$

$$[H_3O^+] \times [OH^-] = 1 \times 10^{-14} \rightarrow [H_3O^+] = \frac{1 \times 10^{-14}}{0.015} = 6.67 \times 10^{-13}$$

$pH = -\log [H_3O^+] = -\log (6.67 \times 10^{-13}) = 12.18$

Alternatively, pH + pOH = 14, thus pH = 14 − 1.82 = 12.18

9.21   Each of the acids in this question is a strong acid, which means that every acid molecule dissociates into an H_3O^+ ion and a negative counter ion. Thus, $[H_3O^+]$ will equal the concentration of the acid. Thus

(a) $pH = -\log[H_3O^+] = -\log(0.0033) = 2.48$

(b) $pH = -\log[H_3O^+] = -\log(0.025) = 1.60$

(c) $pH = -\log[H_3O^+] = -\log(0.00073) = 3.14$

(d) $pH = -\log[H_3O^+] = -\log(0.074) = 1.13$

9.23   A weak acid is one that incompletely dissociates in water. Examples are given in table 9.4, acetic acid and carbonic acid.

9.25   Because the pK_a value of the weak acid, acetic acid, and the pK_b value of the weak base ammonia are approximately equal, they will both be dissociated in aqueous solution to a similar extent. Therefore, the OH^- ions produced by the reaction of ammonia with water will neutralize all of the $[H_3O^+]$ ions created by acetic acid and vice versa, thus the solution will be neutral.

9.27   (a) An acid is a compound that donates a proton, its conjugate base is the compound that is formed upon loss of a hydrogen. Both HNO_2 and H_3O^+ give up a proton to form NO_2^- and H_2O respectively. Therefore HNO_2 and NO_2^- are one conjugate acid-base pair, while H_3O^+ and H_2O are another.

(b) An acid is a compound that donates a proton, its conjugate base is the compound that is formed upon loss of a hydrogen. Both $H_2PO_4^-$ and H_3O^+ give up a proton to form HPO_4^{2-} and H_2O respectively. Therefore $H_2PO_4^-$ and HPO_4^{2-} are one conjugate acid-base pair, while H_3O^+ and H_2O are another.

9.29   The equilibrium constant for the reaction of acetate ion with water is $K_b = 5.75 \times 10^{-10}$.

9.31 $pK_a = -\log(K_a) = -\log(1.74 \times 10^{-5}) = 4.76$

$pK_b = -\log(K_b) = -\log(5.75 \times 10^{-10}) = 9.24$

9.33   From 9.31, $pK_a + pK_b = 4.76 + 9.24 = 14.0$

9.35 $H_2CO_3(aq) + H_2O(l) \leftrightarrow H_3O^+(aq) + HCO_3^-(aq)$

$HCO_3(aq) + H_2O(l) \leftrightarrow H_3O^+(aq) + CO_3^{2-}(aq)$

9.37 $K_{a1} = \dfrac{[H_3O^+][HCO_3^-]}{[H_2CO_3]} = 4.45 \times 10^{-7}$

$$K_{a2} = \frac{[H_3O^+][CO_3^{2-}]}{[HCO_3^-]} = 4.72 \times 10^{-11}$$

9.39   From the equations in 9.37, the amount of H_3O^+ contributed by HCO_3^- is a factor of 10,000 less than that contributed by H_2CO_3 ($10^{-11}/10^{-7} = 1/10,000$). Thus it will be of minor significance.

9.41   Salts composed of cations and anions of strong acids and bases form neutral solutions, while salts having anions of weak acids form basic solutions and salts having cations of weak bases form acidic solutions; thus,

(a) $CaCl_2$ is a salt of a strong base and a strong acid ($Ca(OH)_2$ and HCl), thus the solution is neutral.

(b) Na_2CO_3 is the salt of the anion of a weak acid (H_2CO_3), thus the solution is basic.

(c) $FeCl_3$ is the salt whose cation is a weak acid, thus the solution is acidic.

(d) NH_4Cl is the salt of the cation of a weak base (NH_3), thus the solution is acidic.

(e) $MgSO_4$ is a salt of a strong base and a strong acid ($Mg(OH)_2$ and H_2SO_4), thus the solution is neutral.

9.43   Salts composed of cations and anions of strong acids and strong bases form neutral solutions, while salts having anions of weak acids form basic solutions and salts having cations of weak bases form acidic solutions; thus,

(a) $Ca(NO_3)_2$ is a salt of a strong base and a strong acid ($Ca(OH)_2$ and HNO_3), thus the solution is neutral.

(b) $NaNO_2$ is the salt of the anion of a weak acid (HNO_2), thus the solution is basic.

(c) KCN is the salt of the anion of a weak acid (HCN), thus the solution is basic.

(d) CH_3COONa is the salt of the anion of a weak acid (CH_3COOH), thus the solution is basic.

(e) $(HCOO)_2Mg$ is the salt of the anion of a weak acid ($HCOOH$), thus the solution is basic.

9.45   Because the presence of the conjugate acid allows the solution to neutralize the addition of a base while the presence of the conjugate base allows the solution to neutralize the addition of a base. Because it is a conjugate acid-base pair, the solution remains an a constant pH.

9.47 $pH = pK_a + \log\left(\dfrac{\text{proton acceptor}}{\text{proton donor}}\right) = \log\left(\dfrac{0.0080}{0.0060}\right) = 4.76 + 0.13 = 4.89$

9.49 $pH = pK_a + \log\left(\dfrac{\text{proton acceptor}}{\text{proton donor}}\right) = \log\left(\dfrac{0.075}{0.050}\right) = 7.20 + 0.18 = 7.38$

9.51 $16.2 \text{ mL} \times \dfrac{1 \text{ L}}{1000 \text{ mL}} \times \dfrac{0.0210 \text{ mol}}{\text{L}} = 0.000340 \text{ mol of KOH}$

$0.000340 \text{ mol KOH} = 0.000340 \text{ mol HCl when neutral}$

$\dfrac{0.000340 \text{ mol}}{0.0200 \text{ L}} = 0.0170 \text{ M}$

9.53 $33.2 \text{ mL} \times \dfrac{1 \text{ L}}{1000 \text{ mL}} \times \dfrac{0.0410 \text{ mol}}{\text{L}} = 0.00136 \text{ mol KOH}$

$0.00136 \text{ moles KOH} \times \dfrac{1 \text{ mole } H_2SO_4}{2 \text{ mole KOH}} = 0.000680 \text{ mol } HNO_3 \text{ when neutral}$

$\dfrac{0.000680 \text{ moles}}{0.0400 \text{ L}} = 0.0170 \text{ M}$

9.55 0.500 N solution denotes 0.500 equivalents of acid /liter. An equivalent is defined as the number of moles of dissociable H^+ /mole of acid. In H_3PO_4, 3 H^+ ions can come from a single H_3PO_4 molecule, thus

$\dfrac{0.500 \text{ equivalent}}{\text{L}} = \dfrac{x \text{ mol}}{\text{L}} \times \dfrac{3 \text{ mol } H^+}{\text{mol}}$ $x = 0.167$ moles H_3PO_4 per liter

$0.167 \text{ moles } H_3PO_4 \times 98 \text{ g/mole} = 16.3 \text{ g } H_3PO_4$

Thus, 16.3 g of H_3PO_4 must be dissolved in sufficient water to make 1.00 L of solution.

9.57 $0.024 \text{ M } H_3PO_4 \times \dfrac{3 \text{ mol } H^+}{\text{mol}} = 0.072 \text{ N solution}$

9.59 (a) $0.125 \text{ N } Na^+ \times \dfrac{1 \text{ mol}}{1 \text{ eq of charge}} = 0.125 \text{ M}$

(b) $0.035 \text{ N } K^+ \times \dfrac{1 \text{ mol}}{1 \text{ eq of charge}} = 0.035 \text{ M}$

(c) $0.072 \text{ N } Ca^{2+} \times \dfrac{1 \text{ mol}}{2 \text{ eq of charge}} = 0.036 \text{ M}$

(d) $.028 \text{ N Mg}^{2+} \times \dfrac{1 \text{ mol}}{2 \text{ eq of charge}} = 0.014 \text{ M}$

9.61   (a) 0.0031 M HNO_3 solution. HNO_3 is a strong acid, which means that every HNO_3 molecule dissociates into an H_3O^+ ion and a NO_3^- ion. Thus, $[H_3O^+] = 0.0031$ M.

$$pH = -\log [H_3O^+] = -\log(0.0031) = 2.51$$

(b) 1.0 M HCl solution. HCl is a strong acid, which means that every HCl molecule dissociates into an H_3O^+ ion and a Cl^- ion. Thus, $[H_3O^+] = 1.0$ M.

$$pH = -\log [H_3O^+] = -\log(1.0) = 0.00$$

(c) 0.0069 M HI solution. HI is a strong acid, which means that every HI molecule dissociates into an H_3O^+ ion and an I^- ion. Thus, $[H_3O^+] = 0.0069$ M.

$$pH = -\log [H_3O^+] = -\log(0.0069) = 2.16$$

(d) 0.019 M HBr solution. HNO_3 is a strong acid, which means that every HBr molecule dissociates into an H_3O^+ ion and a Br^- ion. Thus, $[H_3O^+] = 0.0019$ M.

$$pH = -\log [H_3O^+] = -\log(0.019) = 1.72$$

(e) 0.023 M $HClO_3$ solution. $HClO_3$ is a strong acid, which means that every $HClO_3$ molecule dissociates into an H_3O^+ ion and a ClO_3^- ion. Thus, $[H_3O^+] = 0.023$ M.

$$pH = -\log [H_3O^+] = -\log(0.023) = 1.64$$

9.63   (a) 0.0062 M KOH solution. KOH is a strong acid, which means that every KOH molecule dissociates into an OH^- ion and a K^+ ion. Thus, $[OH^-] = 0.0062$ M.

$$pOH = -\log [OH^-] = -\log(0.0062) = 2.21$$

(b) 0.0041 M $Ca(OH)_2$ solution. $Ca(OH)_2$ is a strong acid, which means that every $Ca(OH)_2$ molecule dissociates into two OH^- ions and a Ca^{2+} ion. Thus, $[OH^-] = 0.0082$ M.

$$pOH = -\log [OH^-] = -\log(0.0082) = 2.09$$

(c) 0.028 M $Ba(OH)_2$ solution. $Ba(OH)_2$ is a strong acid, which means that every $Ba(OH)_2$ molecule dissociates into two OH^- ions and a Ba^{2+} ion. Thus, $[OH^-] = 0.056$ M.

$$pOH = -\log [OH^-] = -\log(0.056) = 1.25$$

(d) 1.0 M KOH solution. KOH is a strong acid, which means that every KOH molecule dissociates into an OH^- ion and a K^+ ion. Thus, $[OH^-] = 1.0$ M.

$$\text{pOH} = -\log[\text{OH}^-] = -\log(1.0) = 0.00$$

(e) 0.01 M NaOH solution. NaOH is a strong acid, which means that every NaOH molecule dissociates into an OH^- ion and a Na^+ ion. Thus, $[\text{OH}^-] = 0.010$ M.

$$\text{pOH} = -\log[\text{OH}^-] = -\log(0.010) = 2.00$$

9.65   Salts composed of cations and anions of strong acids and bases form neutral solutions, while salts having anions of weak acids form basic solutions and salts having cations of weak bases form acidic solutions; thus,

(a) $Fe_2(SO_4)_3$ is a compound whose cation is a weak acid, thus the solution is acidic.

(b) NaBr is a salt of a strong base and a strong acid (NaOH and HBr), thus the solution is neutral.

(c) $NaNO_2$ is the salt of the anion of a weak acid (HNO_2), thus the solution is basic.

(d) NH_4NO_3 is the salt of the cation of a weak base, thus the solution is acidic.

(e) $Mg(CN)_2$ is the salt of the anion of a weak acid (HCN), thus the solution is basic.

9.67   The combination of a weak acid and the salt of its conjugate base is a buffer solution. The pH of a buffer is

$$\text{pH} = \text{p}K_a + \log\left(\frac{\text{proton acceptor}}{\text{proton donor}}\right) = 8.2 = 7.2 + x$$

$$x = 1.0 = \log\left(\frac{\text{proton acceptor}}{\text{proton donor}}\right) \rightarrow HPO_4^{2-}/H_2PO_4^- = 10/1$$

9.69 $62.0 \text{ mL} \times \dfrac{1 \text{ L}}{1000 \text{ mL}} \times \dfrac{0.0250 \text{ mol}}{\text{L}} = 0.00155 \text{ mol HCl}$

$0.00155 \text{ mol HCl} \times \dfrac{1 \text{ mol Ca(OH)}_2}{2 \text{ mol HCl}} = 0.000775 \text{ mol Ca(OH)}_2$ when neutral

$\dfrac{0.000775 \text{ mol}}{0.0250 \text{ L}} = 0.0310 \text{ M}$

9.71   (a) HCl is a strong acid, which means that every HCl molecule dissociates into an H_3O^+ ion and a Cl^- ion. Thus, $[H_3O^+] = [\text{HCl}] = .0034$ M. It is also known that

$$[H_3O^+] \times [\text{OH}^-] = 1.00 \times 10^{-14} \rightarrow [\text{OH}^-] = \frac{1.00 \times 10^{-14}}{0.0034} = 3.0 \times 10^{-12}$$

$$pOH = -\log[OH^-] = 11.53$$

(b) HNO$_3$ is a strong acid, which means that every HNO$_3$ molecule dissociates into an H$_3$O$^+$ ion and a NO$_3^-$ ion. Thus, [H$_3$O$^+$] = [HNO$_3$] = 0.025 M. It is also known that

$$[H_3O^+] \times [OH^-] = 1.00 \times 10^{-14} \rightarrow [OH^-] = \frac{1.00 \times 10^{-14}}{0.025} = 4.0 \times 10^{-13}$$

$$pOH = -\log[OH^-] = 12.40$$

(c) HBr is a strong acid, which means that every HBr molecule dissociates into an H$_3$O$^+$ ion and a Br$^-$ ion. Thus, [H$_3$O$^+$] = [HBr] = 0.00073 M. It is also known that

$$[H_3O^+] \times [OH^-] = 1.00 \times 10^{-14} \rightarrow [OH^-] = \frac{1.00 \times 10^{-14}}{0.00073} = 1.4 \times 10^{-11}$$

$$pOH = -\log[OH^-] = 10.86$$

(d) HI is a strong acid, which means that every HI molecule dissociates into an H$_3$O$^+$ ion and an I$^-$ ion. Thus, [H$_3$O$^+$] = [HI] = 0.074 M. It is also known that

$$[H_3O^+] \times [OH^-] = 1.00 \times 10^{-14} \rightarrow [OH^-] = \frac{1.00 \times 10^{-14}}{0.074} = 1.4 \times 10^{-13}$$

$$pOH = -\log[OH^-] = 12.87$$

9.73   Ammonium acetate is a salt that will dissociate into an NH$_4^+$ ion and an acetate ion, CH$_3$COO$^-$. NH$_4^+$ and CH$_3$COO$^-$ ions can interact with water as an acid and a base, respectively. Inspection of Table 9.5 shows that the pK_a of NH$_4^+$ and the pK_b of CH$_3$COO$^-$ are the same, 9.24. This signifies that the amount of OH$^-$ and H$_3$O$^+$ ions produced by the NH$_4^+$ and CH$_3$COO$^-$ ions will be equal. Therefore the solution will be neutral.

9.75   A buffer approximately equal amounts of a weak acid and a salt of its conjugate base.

9.77   A conjugate acid-base pair is a weak acid and the basic anion that results from its dissociation.

9.79   pH = $-\log$[H$_3$O$^+$] = 1.50 $\rightarrow$ [H$_3$O$^+$] = 0.0316 mol/liter

0.0316 mol/liter × 2.00 liter = 0.0632 mol of H$_3$O$^+$

To neutralize, 0.0632 mol of HCO$_3^-$ must be added.

0.0632 mol × 84.0 g/mol = 5.31 g of NaCO$_3$ is needed.

9.81   pH = 8.38

9.83   0.15 atm

9.85   When 50% of the acid has been neutralized, the concentration of acid (e.g., acetic acid) in solution is equal to the concentration of acid anion (e.g., acetate).

According to the Henderson-Hasselbalch equation, the logarithm is then equal to zero, and the pH is equal to the pK_a.

9.87 $Na_3PO_4 > Na_2HPO_4 > NaH_2PO_4$

9.89 $CO_3^{2-} > HCO_3^-$; $PO_4^{3-} > HPO_4^{2-} > H_2PO_4^-$

9.91   A basic solution will turn bromcresol green to blue, while an acidic solution will turn it yellow. As a pH of 2 is acidic and a pH of 12 is basic, the pH of solution A is 2, and the pH of solution B is 12.

Chapter 10

Chemical and Biological Effects of Radiation

10.1 α radiation is the emission of α particles that are helium nuclei, with a charge of +2 and a mass of 4 amu.

10.3 γ radiation has no mass or charge, but can penetrate very deep into solids.

10.5   The sum of the mass numbers of the particles on the left must equal the mass numbers of the particles on the right; $240 + x = 243 + 1$, $x = 4$. Similarly, the atomic numbers of the particles on the left must equal the atomic numbers of the particles on the right; $95 + y = 97 + 0$, $y = 2$. Thus the missing component must be a particle with a mass number of 4 and an atomic number of 2. This is an α-particle, ^4_2He.

10.7   The sum of the mass numbers of the particles on the left must equal the mass numbers of the particles on the right; $40 = 40 + 0$, $x = 0$. Similarly, the atomic numbers of the particles on the left must equal the atomic numbers of the particles on the right; $19 = 20 + y$, $y = -1$. Thus the missing component must be a particle with a mass number of 0 and an atomic number of -1. This is an electron, $^0_{-1}\text{e}$.

10.9   Induced radioactivity results when a nonradioactive element is bombarded with high-energy subatomic particles.

10.11   A radioactive decay series is a series of nuclear reactions that begins with an unstable nucleus and ends with the formation of a stable isotope.

10.13 $\dfrac{N}{N_0} = \left(\dfrac{1}{2}\right)^x$, where x is the number of half-lives of C-14 that have passed.

$\dfrac{N}{N_0} = 0.233$, thus $0.233 = (1/2)^x$ $\rightarrow$ $\log(0.233) = x \log(1/2)$ $\rightarrow$ $x = \log(0.233)/\log(1/2) =$ 2.10 half-lives × 5730 years/half-life = 1.20×10^4 years.

10.15 β-radiation passes through solids more easily than α-radiation, thus β-radiation.

10.17   As an X-ray passes through tissue, it interacts with the tissue, loses energy, and its intensity is lowered.

10.19 γ-radiation will ionize water molecules, causing the water molecules to lose an electron and produce hydrated electrons.

10.21   The secondary chemical processes have the most long-term consequences because the free radicals that are formed can undergo harmful reactions with biomolecules.

10.23   Radiation sickness is the result of a non-lethal exposure to radiation. It is characterized by nausea, lethargy, and a drop in the white blood cell count.

10.25   The intensity of radiation decreases by a factor that is inversely proportional to the square of the distance. Thus, if the technician doubles the distance between herself and the radiation, the intensity will decrease by a factor of 4. Therefore, she must move from 4 feet away to 8 feet away.

10.27   The Geiger-Muller counter actually measures the amount of ions that are produced due to the presence of the radiation.

10.29   A rad is the amount of energy that a sample absorbs when it is exposed to radiation. This factor does not account for effects that the radiation may have on the sample.

10.31   No, α-particles are more damaging to tissue (RBE = 10) than γ-particles (RBE = 1.0) and thus 1 rad of α-particles will be more harmful than 1 rad of γ-particles.

10.33   No, in order to be used for diagnosis, the radiation must be detected outside the body. α-particles do not penetrate tissue and thus would remain in the body and not be detected externally.

10.35   Yes, cobalt-60 undergoes radioactive decay to emit γ-radiation that can easily penetrate tissue to cause radiation damage to the cancerous area.

10.37   No, positrons can only be detected when they combine with electrons to form γ-rays, which in turn are only detected when they interact with matter to form ions.

10.39   No, in order to be used for diagnosis, the radiation must be detected outside the body. The β-particles that result from iodine-131 do not penetrate tissue and thus would remain in the body and cannot be detected externally.

10.41   A CAT scan produces images of the internal structures in the body with the use of X-rays.

10.43   30 hours = 5 half-lives. $\dfrac{N}{N_0} = \left(\dfrac{1}{2}\right)^x$, where $x = 5$ → $(1/2)^5 = 0.03125$, thus 0.03125×1 g $= 0.03125$ g $= 31.25$ mg remains.

10.45 $6.3\,\text{mCi} \times \dfrac{\text{Ci}}{1000 \text{ mCi}} \times \dfrac{3.7 \times 10^{10} \text{ dps}}{\text{Ci}} = 2.3 \times 10^8$ disintegrations per second.

10.47   4 hours = 240 minutes. Since 110 minutes = 1 half-life, 240 minutes = 2.18 half-lives.

$\dfrac{N}{N_0} = \left(\dfrac{1}{2}\right)^x$, where $x = 2.18$ → $(1/2)^{2.18} = 0.221$, thus there is 22.1% of the original sample left after 4 hours.

10.49   Radioactivity is the name of the process where atomic nuclei spontaneously decompose.

10.51   No, γ-radiation does not alter the atomic mass or number of an atom that it interacts with.

10.53   Any tissue that is rapidly dividing such that any damage that may result from an interaction will affect the ability of the tissue to perform its physiological function.

10.55   In nuclear fission, unstable nuclei of high atomic mass are split by an interaction with a high-energy particle into nuclei with lower atomic mass. In nuclear fusion, nuclei with lower atomic mass fuse together to form new nuclei of higher atomic mass. In either case, an enormous amount of energy is released.

10.57   X-rays are produced by electrons accelerated from an anode at high voltage and then colliding with a heavy metal cathode.

10.59   No. K-shell X-rays are the result of ejection of an electron from an inner atomic electron shell. This is followed by an outer shell electron filling the vacancy. The difference in energy is lost as an X-ray of frequency characteristic of the metal target. Bremsstrahlung are X-rays generated by close encounters of incoming electrons with atomic nuclei of the target. The electrons are slowed by attraction to the highly charged metal nucleus and therefore lose energy, which takes the form of X-rays.

10.61   X-rays passing through tissue encounter regions of differing density, and are therefore absorbed differentially. The emerging pattern of differential absorption reveals the underlying structure.

10.63   Stratospheric ozone is formed when oxygen is exposed to high-energy cosmic radiation. This cleaves O_2 into free radicals that can recombine into O_3 (ozone) molecules.

Chapter 11

Saturated Hydrocarbons

11.1 The number of covalent bonds formed by the atom of an element in forming a compound.

11.3 In the molecular formula, the carbons must have four bonds and the hydrogens must have one bond. The only formula that obeys this rule is (a)

11.5 The molecular formula must be a structure that has all of the atoms attaining the correct combining power. The only formula where this is true is (b)

11.7 (a)

(b)

C—C & C—H

(c)

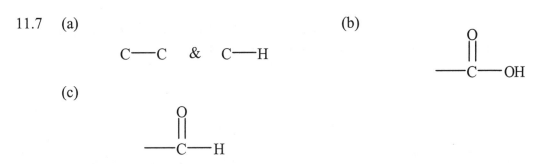

11.9 (a) alcohol   (b) alkene   (c) ketone   (d) aromatic

(e) ketone   (f) ether   (g) carboxylic acid   (h) amide

11.11 A carbon atom in an alkane can form four equivalent bonds by sharing electrons with other atoms, therefore the four electrons in the outer shell of carbon in the ground state must be equal. However, the four electrons in the outer shell of carbon in the ground state are not equal (one is in the s orbital, three are in p orbitals). This discrepancy is corrected by proposing that the four electrons fill four equivalent sp^3 orbitals, which are formed by combining (hybridizing) the three p and one s orbitals. Then, the four electrons can form four equivalent bonds.

11.13 The expanded structure explicitly shows each C–C and C–H bond of an alkane, therefore the answers are:

(a)   (b)

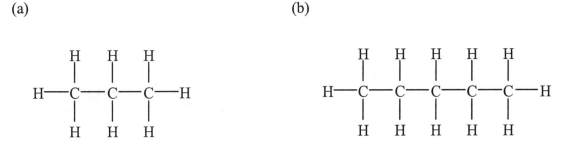

(c)

$$\begin{array}{ccccc} & H & H & H & H & H \\ & | & | & | & | & | \\ H\!-\!C\!-\!C\!-\!C\!-\!C\!-\!C\!-\!H \\ & | & | & | & | & | \\ & H & H & H & C & H \\ & & & & \diagup | \diagdown \\ & & & H \; H \; H \end{array}$$

11.15   The line structural formulas only show lines to denote the C–C bonds. Bonds are put in to clearly denote where atoms exist. Therefore, the answers are:

(a)   (b)

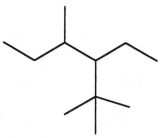

(c)   (d)

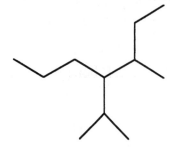

11.17   (a) Both compounds have a molecular formula of C_5H_{12}, therefore they are either the same compound or constitutional isomers. By drawing the expanded structural formula of the first compound, it is obvious that they are the same compound.

(b) Both compounds have a molecular formula of C_6H_{14}, therefore they are either the same compound or constitutional isomers. By drawing the expanded structural formula of the first compound, it is seen that they are not the same compound, therefore they are constitutional isomers.

(c) Both compounds have a molecular formula of C_6H_{14}, therefore they are either the same compound or constitutional isomers. By drawing the expanded structural formula of the second compound, it is obvious that they are not the same compound, therefore they are constitutional isomers.

11.19

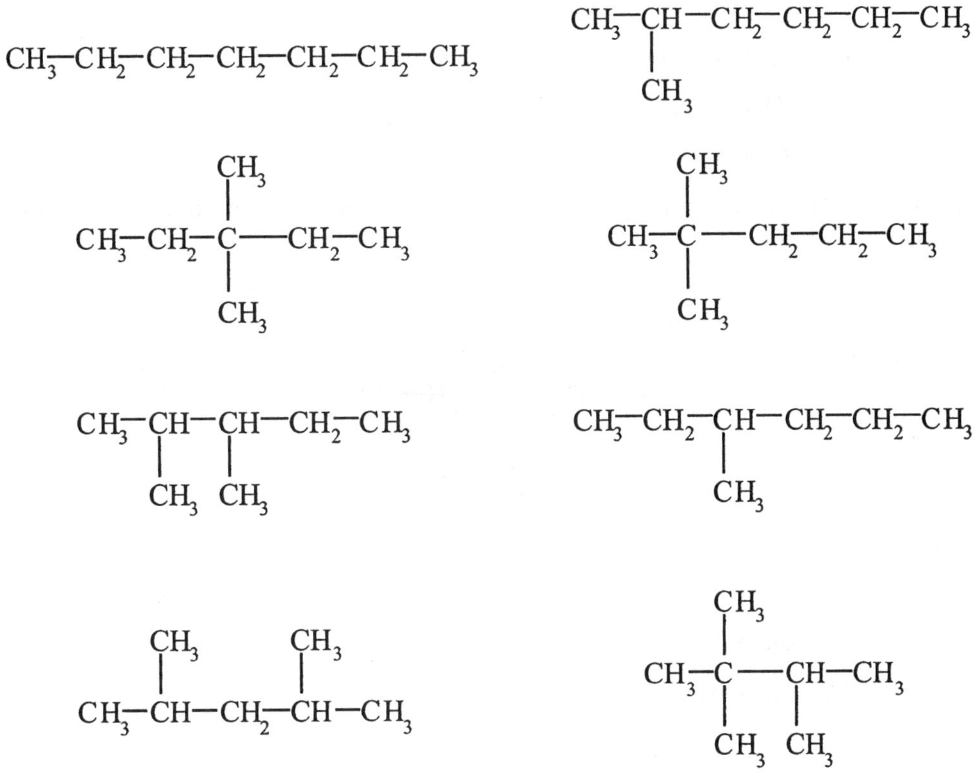

$CH_3-CH_2-CH_2-CH_2-CH_2-CH_2-CH_3$

$CH_3-CH-CH_2-CH_2-CH_2-CH_3$
$\qquad\quad|$
$\qquad\quad CH_3$

$\qquad\quad CH_3$
$\qquad\quad\ |$
$CH_3-CH_2-C-CH_2-CH_3$
$\qquad\quad\ |$
$\qquad\quad CH_3$

$\qquad\quad CH_3$
$\qquad\quad\ |$
$CH_3-C-CH_2-CH_2-CH_3$
$\qquad\quad\ |$
$\qquad\quad CH_3$

$CH_3-CH-CH-CH_2-CH_3$
$\qquad\ \ |\quad\ |$
$\qquad\ CH_3\ CH_3$

$CH_3-CH_2-CH-CH_2-CH_2-CH_3$
$\qquad\qquad\ \ |$
$\qquad\qquad\ CH_3$

$\ \ CH_3\qquad\quad CH_3$
$\quad\ |\qquad\qquad\ |$
$CH_3-CH-CH_2-CH-CH_3$

$\qquad\qquad CH_3$
$\qquad\qquad\ |$
$CH_3-C-CH-CH_3$
$\qquad\quad\ |\quad\ |$
$\qquad\ CH_3\ CH_3$

$(CH_3-CH_2)_3CH$

11.21   Method for naming alkanes:

(i)   Find longest carbon chain. Form base name accordingly.

(ii)   Find and name groups that are attached to this primary chain.

(iii) Put names of these groups in alphabetical order at beginning of name. Include numbers to denote which carbon(s) the groups are attached to. Number these groups such that the numbering results in using lowest possible numbers.

(a)   (i) 4 carbons = butane

(ii) 1 CH_3 group = methyl attached

(iii) 2-methylbutane

(b) (i) 5 carbons = pentane

(ii) 1 CH_3 group = methyl attached

(iii) 3-methylpentane

(c) (i) 6 carbons = hexane

(ii) 3 CH_3 groups = methyl attached

(iii) 2,2,4-trimethylhexane

(d) (i) 6 carbons = hexane

(ii) 1 CH_3 group = methyl

& 1 CH_3CHCH_3 = isopropyl group

(iii) 3-isopropyl-4-methylhexane

This compound is interesting in that, if pictured as the structure below, it can also be named as follows:

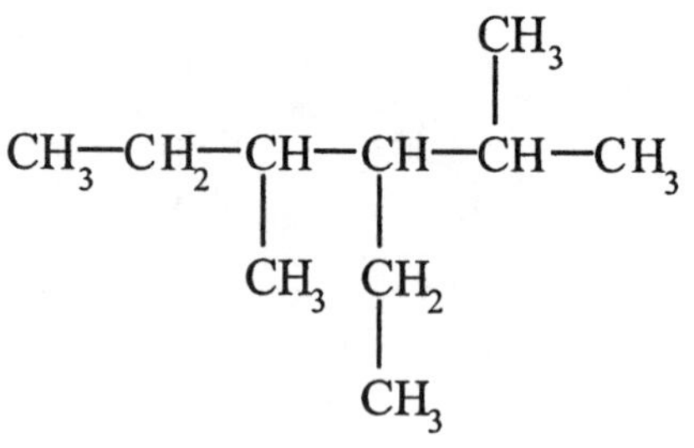

$$CH_3-CH_2-CH-CH-CH-CH_3$$

with CH_3 above the fourth carbon, CH_3 and CH_2 below the third and fourth carbons, and CH_3 below the CH_2.

(i) 6 carbons = hexane

(ii) 2 CH_3 group2 = methyl

& 1 CH_2CH_3 = ethyl group

(iii) 3-ethyl-2,4 dimethylhexane

11.23 Method for drawing structures from alkane names:

(i)   Draw longest carbon chain from the base name.

(ii) Add side groups on appropriate carbons along chain.

(iii) Fill in hydrogens such that each carbon has four bonds.

(a) (i) butane = 4

C–C–C–C

(ii) Methyl = CH_3 group on the second carbon

(iii)

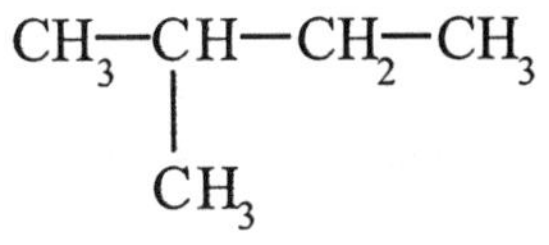

$$CH_3-CH-CH_2-CH_3$$
$$|$$
$$CH_3$$

(b)   (i) hexane = 6 carbons

C–C–C–C–C-C

(ii) Methyl groups = CH_3 on C-2, C-2, and C-4

(iii)

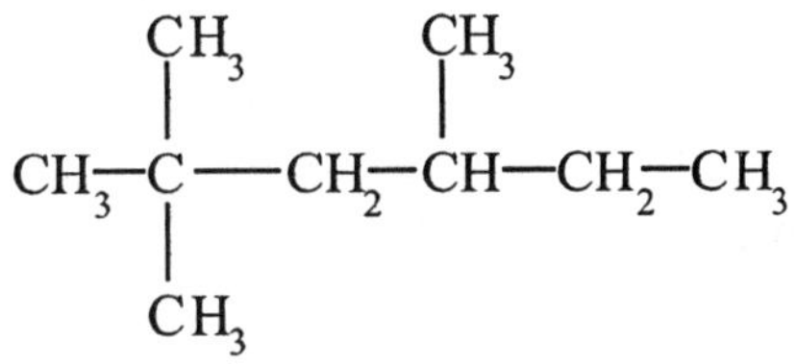

$$\begin{array}{ccc} CH_3 & & CH_3 \\ | & & | \\ CH_3-C-CH_2-CH-CH_2-CH_3 \\ | \\ CH_3 \end{array}$$

(c)   (i) hexane = 6 carbons

C–C–C–C–C-C

(ii) Methyl groups = CH_3 on C-2, and C-3; isopropyl = CH_3CHCH_3 on C-3

(iii)

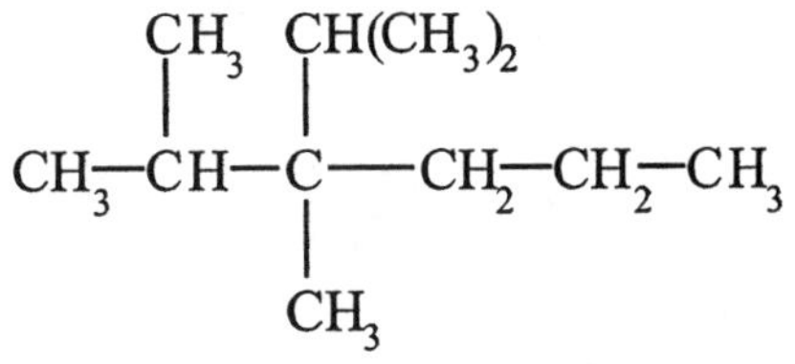

$$\begin{array}{cc} CH_3 & CH(CH_3)_2 \\ | & | \\ CH_3-CH-C-CH_2-CH_2-CH_3 \\ | \\ CH_3 \end{array}$$

(d)   (i) Octane = 8 carbons

C–C–C–C–C-C–C–C

(ii) Methyl groups = $-CH_3$ on C-2, and C-2; isopropyl = CH_3CHCH_3 on C-4; ethyl group = $-CH_2CH_3$ on C-4.

(iii)

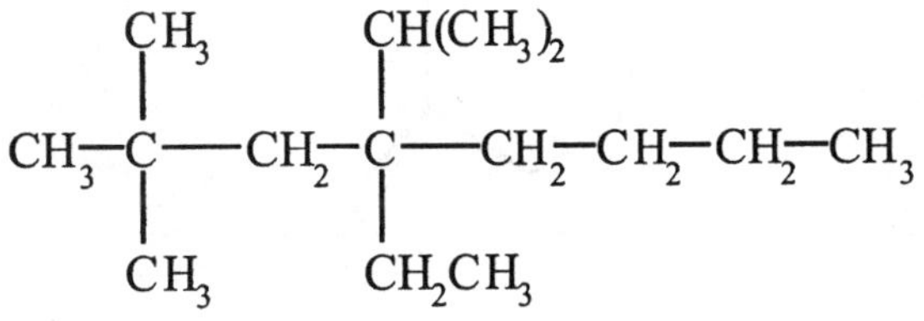

$$\begin{array}{cc} CH_3 & CH(CH_3)_2 \\ | & | \\ CH_3-C-CH_2-C-CH_2-CH_2-CH_2-CH_3 \\ | & | \\ CH_3 & CH_2CH_3 \end{array}$$

(e)   (i) heptane = 7 carbons

C–C–C–C–C–C–C

(ii) Methyl group = –CH_3 on C-2; sec-butyl = $CH_3CH_2CHCH_3$ on C-4

(iii)

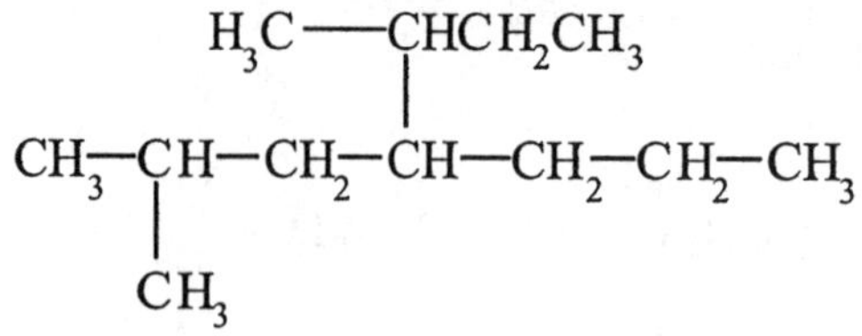

11.25   (a) Octane = 8 carbons, Thus:

(b) Cyclohexane = 6 carbons.

Choose a carbon for one of the methyl groups, label that carbon in the ring as C-1.

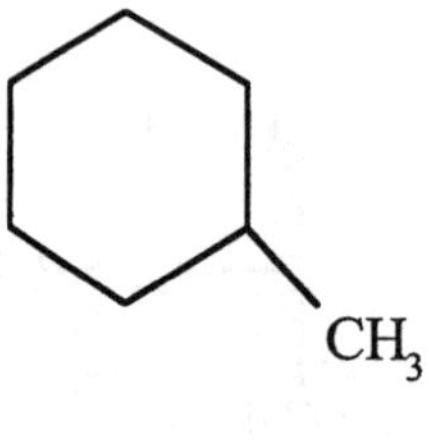

Place other groups (methyl and isopropyl) on the correct carbons by counting clockwise around ring from C-1.

CH(CH$_3$)$_2$

4

CH$_3$

1

CH$_3$

2

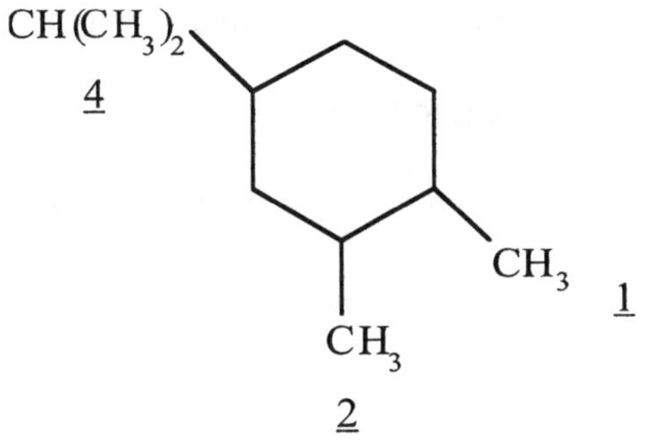

(c) Cyclohexane = 6 carbons.

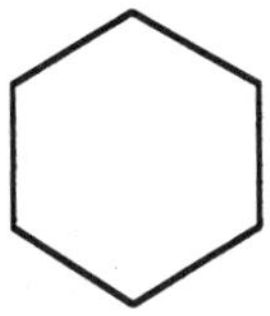

Choose a carbon for t-butyl group, label that carbon in the ring as C-1.

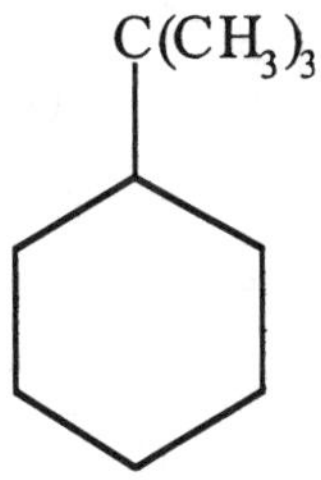

C(CH$_3$)$_3$

Place other group (ethyl) on the correct carbon (C-3) by counting clockwise around ring from C-1.

C(CH$_3$)$_3$

CH$_2$CH$_3$

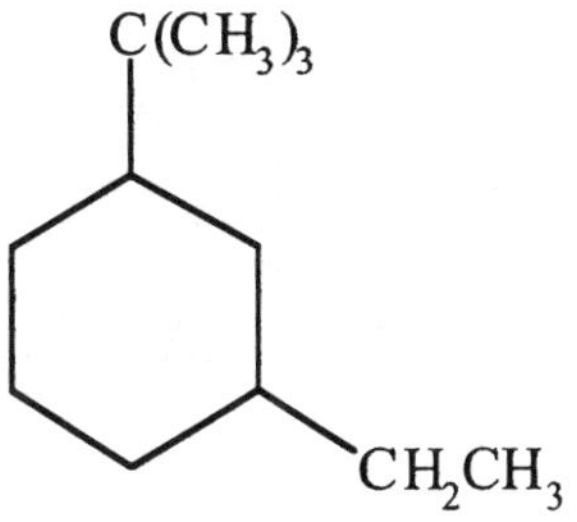

11.27 1° - primary - carbons are connected to one other carbon.

2° - secondary - carbons are connected to two other carbons.

3° - tertiary - carbons are connected to three other carbons.

4° - quaternary - carbons are connected to four other carbons.

(a)

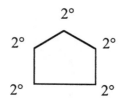

(b)

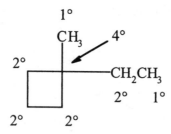

(c)

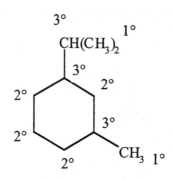

11.29 (a) One methyl (CH_3) group cannot be part of a ring, therefore it must be connected to the ring. This leaves C_5H_9 to form the (cyclopentane) ring. The compound is methylcyclopentane.

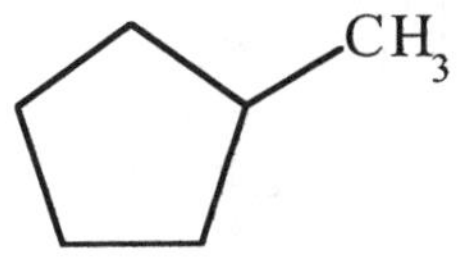

(b) An ethyl group = C_2H_5, which cannot be part of the ring structure, thus C_4H_7 remains to form the ring. Therefore the ring is cyclobutane. The compound is thus ethylcyclobutane.

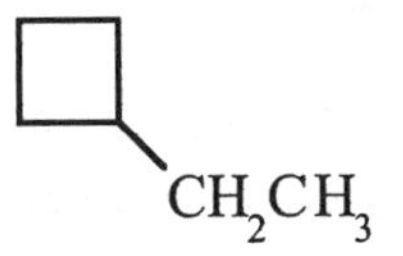

11.31 To exist as cis-trans isomers in cycloalkanes, there must exist two carbons within the ring, each of which contains two different substituents.

(a) No cis-trans isomerism, because only one carbon has two different substituents on it.

(b) Both C-1 and C-2 have two different substituents, therefore this compound does exhibit cis-trans isomers as:

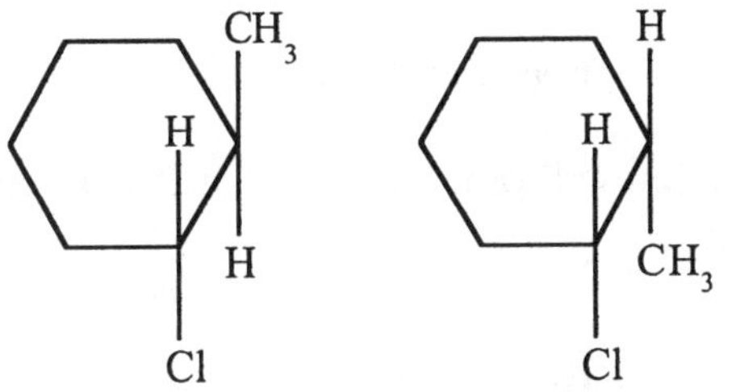

(c) Both C-1 and C-3 have two different substituents, therefore this compound does exhibit cis-trans isomers as

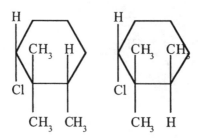

11.33 (a) Hexane has the higher boiling point because it has the higher molecular mass and, therefore, greater secondary forces.

(b) Cyclohexane has the higher boiling point because it has the higher molecular mass and, therefore, has greater secondary forces.

(c) Methylcyclohexane has the higher boiling point because it has the higher molecular mass and, therefore, has greater secondary forces.

(d) Hexane has the higher boiling point because it is not branched and, therefore, has greater secondary forces.

(e) Cyclopentane has the higher boiling point because it is cyclic, packs tightly, and, therefore, has greater secondary forces.

11.35 Alkanes (and cycloalkanes) are very inert substances, which do not undergo many reactions. The only two common reactions that they undergo are (i) combustion with oxygen and (ii) halogenation = substitution of a hydrogen with a halogen (F, Cl, Br, etc...) in the presence of UV or heat.

Thus, there is no reaction for a, b, c, and d.

(e) This halogenation reaction will proceed where a Cl will replace one of the hydrogens on the butane to form:

$$CH_3-CH-CH_2-CH_3 \quad\quad or \quad\quad CH_2-CH_2-CH_2-CH_3$$
$$\quad\quad\;\; | \quad\quad\quad\quad\quad\quad\quad\quad\quad\quad\quad | $$
$$\quad\quad\;\; Cl \quad\quad\quad\quad\quad\quad\quad\quad\quad\quad Cl$$

11.37 Complete combustion of an alkane, C_nH_m is given by the equation

$$C_nH_m + z\,O_2 \rightarrow n\,CO_2 + (m/2)\,H_2O$$

Where $z = (n + m/2)$

(a) pentane $= C_5H_{12}$, thus $n = 5$, $m = 12$, $z = 8$.

$$C_5H_{12} + 8\,O_2 \rightarrow 5\,CO_2 + 6\,H_2O$$

(b) 2-methylbutane $= C_5H_{12}$, thus $n = 5$, $m = 12$, $z = 8$.

$$C_5H_{12} + 8\,O_2 \rightarrow 5\,CO_2 + 6\,H_2O$$

(c) cyclohexane $= C_6H_{12}$, thus $n = 6$, $m = 12$, $z = 9$.

$$C_6H_{12} + 9\,O_2 \rightarrow 6\,CO_2 + 6\,H_2O$$

(d) methylcyclopentane $= C_6H_{12}$, thus $n = 6$, $m = 12$, $z = 9$.

$$C_6H_{12} + 9\,O_2 \rightarrow 6\,CO_2 + 6\,H_2O$$

11.39 One carbon can bond to one or more other carbon atoms in many different patterns.

11.41

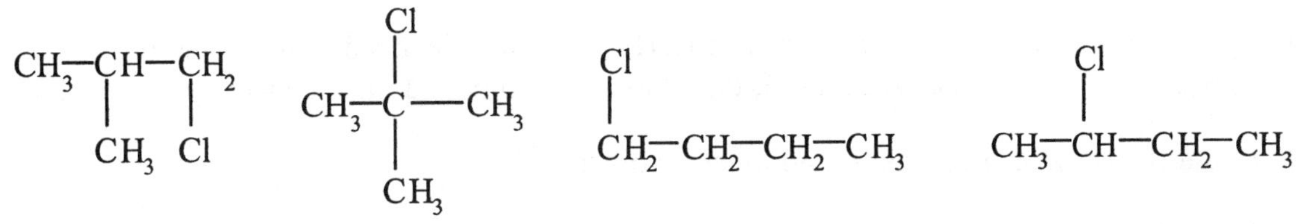

11.43 (a) As the first compound has one chloro group while the second has 2 Cl, these two structures must be different compounds that are not isomers.

(b) As the first compound is a cycloalkane and the second is not, these two structures must be different compounds that are not isomers.

(c) Both compounds have a molecular formula of C_9H_{18}; therefore they are either the same compound or isomers. As the first compound has a five-membered ring while the second has a six-membered ring, the two compounds must be constitutional isomers.

(d) Both compounds have a molecular formula of C_9H_{18}; therefore they are either the same compound or isomers. As the first compound has both a methyl and ethyl group connected to the ring, while the second has a single propyl group connected to the ring, they must be constitutional isomers.

(e) The molecular formula of the first compound is C_9H_{18} while that of the second is C_9H_{20}; these two structures must be different compounds that are not isomers.

(f) Both compounds have a molecular formula of $C_4H_{11}Cl$; therefore they are either the same compound or isomers. As the first compound has a tertiary carbon while the second does not, the two compounds must be constitutional isomers.

(g) Both compounds have a molecular formula of $C_4H_{11}Cl$; therefore they are either the same compound or isomers. As the first compound has a quartenary carbon while the second does not, the two compounds must be constitutional isomers.

(h) Both compounds have a molecular formula of $C_6H_{10}Cl_2$; therefore they are either the same compound or isomers. As the first compound has both hydrogen and chloro groups on the same side of the ring, while the second has them on the opposite sides, they must be cis-trans isomers.

(i) Both compounds have a molecular formula of $C_6H_{10}Cl_2$; therefore they are either the same compound or isomers. As the first compound has the chloro groups on C-1 and C-3 while the second compound has them on C-1 and C-2, they must be constitutional isomers.

(j) Both compounds have a molecular formula of $C_4H_8Cl_2$; therefore they are either the same compound or isomers. Expansion of the first compound shows that the two structures are the same compound.

11.45 Cyclic alkanes have the general formula C_nH_{2n} while acyclic alkanes have the formula C_nH_{2n+2}. $C_5H_{10} = C_nH_{2n}$ when $n = 5$, therefore this compound must be a cyclic alkane.

11.47 Count the carbons, hydrogens, and other atoms to get:

(a) C_6H_{14} (b) C_8H_{18} (c) $C_6H_{11}Cl$ (d) $C_6H_{11}Cl$

(e) C_9H_{20} (f) C_6H_{12}

11.49 (a) Complete combustion of an alkane, C_nH_m is given by the equation

$$C_nH_m + z\ O_2 \rightarrow n\ CO_2 + (m/2)\ H_2O$$

$$\text{Where } z = (n + m/2)$$

$$C_6H_{14} + 19/2\ O_2 \rightarrow 6\ CO_2 + 7\ H_2O$$

The text states that coefficients should be whole numbers, so multiply all coefficients by 2 to get:

$$2C_6H_{14} + 19\ O_2 \rightarrow 12\ CO_2 + 14\ H_2O$$

(b) In a monobromination reaction, the bromine can replace any hydrogen to form different isomers:

```
                              CH₃  CH₃
                               |    |
CH₃-CH-CH-CH₂-Br          CH₃-C----CH-CH₃
     |   |                     |
    CH₃ CH₃                    Br
```

11.51 $CH_3\text{-}CH_2\text{-}CH_2\text{-}CHCl_2$ & $CH_3\text{-}CH_2\text{-}CCl_2\text{-}CH_3$ & $CH_2Cl\text{-}CHCl\text{-}CH_2\text{-}CH_3$ &
$CH_2Cl\text{-}CH_2\text{-}CHCl\text{-}CH_3$ & $CH_2Cl\text{-}CH_2\text{-}CH_2\text{-}CH_2Cl$

11.53 The alkanes in oil possess only the weak London attractive forces, whereas water contains the much stronger hydrogen bonding forces. Alkane molecules are much more weakly attracted to each other than water molecules and this results in a lower density for oil.

11.55 Chemical reactions are used to convert alkanes to members of other families.

11.57 The molecules of lipstick and petroleum jelly attract each other because they are both hydrocarbons with the same type of secondary force (London). Therefore petroleum jelly dissolves the lipstick while H_2O will not.

11.59 160.43 g CH_4 × 1 mol/16.04 g = 10.00 mol of CH_4

10.00 mole CH_4 × 803 kJ/1 mol CH_4 = 8.03×10^3 kJ.

11.61 In incomplete combustion, the products are CO and H_2O, thus the reaction of hexane (C_6H_{14}) with O_2 to form CO and water must be balanced:

$$2\ C_6H_{14} + 13\ O_2 \rightarrow 12\ CO + 14\ H_2O$$

11.63 The ozone layer is located in the stratosphere, a middle layer of atmosphere, and protects plant and animal life by absorbing much of the cancer causing ultraviolet radiation from space. Any decrease in the amount of ozone will let more UV radiation strike the Earth's surface and will have a detrimental effect on life. Various aerosol can propellants and refrigeration and air conditioning coolants escape from earth and enter the atmosphere, where they can react with ozone and decrease its concentration in the atmosphere.

11.65 There are negligible attractive secondary forces between the highly polar water and nonpolar alkane coatings. The result is that water does not diffuse through the nonpolar coatings and remains inside the fruit or vegetable.

11.67 Complete combustion yields CO_2 and water, whereas incomplete combustion yields CO and water. The amount of water produced is the same in both reactions. Incomplete combustion uses less O_2, while the moles of CO_2 and CO produced are the same. Incomplete combustion produces less heat because the carbon atoms are not oxidized to their fully oxidized state. Less heat translates to less power in the engine and lower miles per gallon.

11.69   Electrical conduction requires ions. NaCl is an ionic compound and the sodium and chloride ions conduct electricity both when NaCl is in solution or in the molten state. Organic compounds with rare exception are covalent, not ionic, and the absence of ions in these compounds renders them unable to conduct electricity.

11.71 $\dfrac{1-\text{Chloropropane}}{2-\text{Chloropropane}}$ = reactivity ratio of two products × ratio of number of hydrogens

$$= (1/4) \times (6/2) = \text{¾}$$

% 1-chloropropane = $(3/7) \times 100 = 42.8\%$

% 2-chloropropane = $(4/7) \times 100 = 57.2\%$

11.73   The molecular masses of methane and propane are 16.04 and 44.09, respectively.

$$\dfrac{213\,\text{kcal}}{\text{mol}} \times \dfrac{\text{mol}}{16.04\,\text{g}} = 13.3 \text{ kcal/g for methane}$$

$$\dfrac{531\,\text{kcal}}{\text{mol}} \times \dfrac{\text{mol}}{44.09\,\text{g}} = 12.0 \text{ kcal/g for propane}$$

11.75

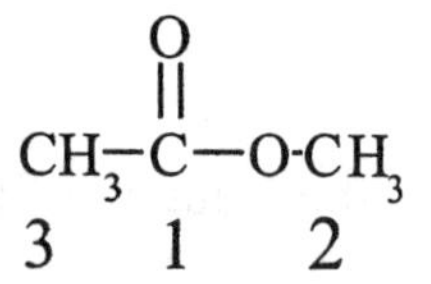

$$\underset{3 \quad\ \ 1 \quad\ \ 2}{\text{CH}_3\text{—C—O-CH}_3}$$

with O double-bonded to the central carbon (C=O).

Chapter 12

Unsaturated Hydrocarbons

12.1   An alkane has the formula C_nH_{2n+2}, while alkenes or cycloalkanes have the formula C_nH_{2n}. If $n = 5$, then the formula of an alkane will be C_5H_{12}.

12.3   An alkane has the formula C_nH_{2n+2}, while alkenes or cycloalkanes have the formula C_nH_{2n}. A single hydrogen in these formulae can also be exchanged with a halogen (Group VII) atom. If a molecular formula of a compound does not follow these guidelines, a compound where all of the carbon atoms have four bonds cannot be created. Therefore, (a) is incorrect while (b) and (c) are correct.

12.5   Counting the number of carbons and hydrogens gives (a) C_5H_{12} and (b) C_7H_{12}. In (b) note that the carbon that has a CH_3 connected to it does not have a hydrogen.

12.7 sp^2 hybridization means that the carbon atom has three equivalent orbitals and one remaining p orbital. This hybridization is used in the formation of carbon-carbon double bonds and thus those carbons in these compounds that are part of a double bond have sp^2 hybridization.

     Therefore, carbons 3, 4, 8, and 9 have sp^2 hybridization.

12.9

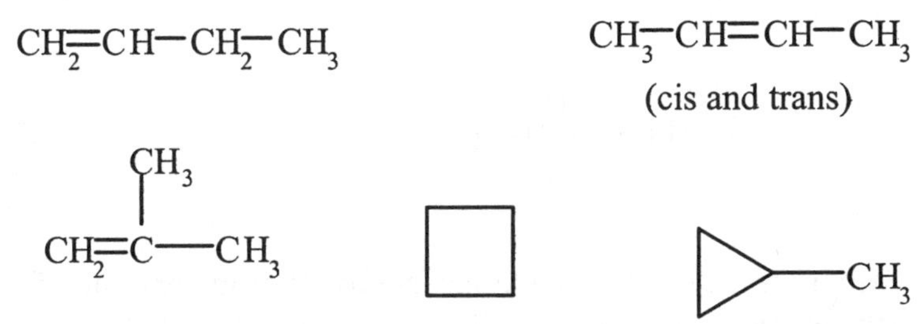

$$CH_2{=}CH{-}CH_2{-}CH_3 \qquad\qquad CH_3{-}CH{=}CH{-}CH_3$$
$$\text{(cis and trans)}$$

12.11 (a) C-C-C-C-C is the skeleton. One of the bonds between carbons must be a double bond, therefore possibilities include:

     C=C-C-C-C   or   C-C=C-C-C.

     Filling in the hydrogens gives:

     $CH_2{=}CH{-}CH_2{-}CH_2{-}CH_3$ and $CH_3{-}CH{=}CH{-}CH_2{-}CH_2$ (cis or trans)

(b) Methyl groups (CH_3) must be on the end of a chain and cannot be part of a double bond. Therefore, the skeleton must be

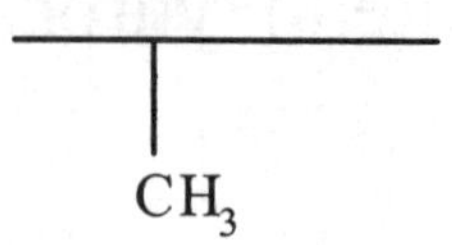

$$
\begin{array}{c}
\text{———————} \\
| \\
CH_3
\end{array}
$$

And the carbon-carbon double bond must be between the second and third carbons of the main chain. Putting this double bond in and filling in the hydrogens gives:

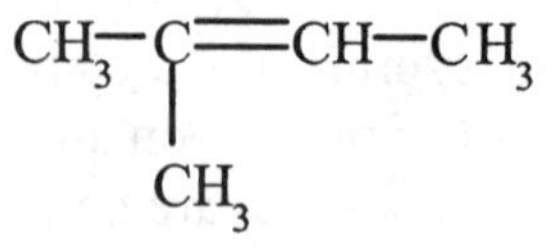

$$
\begin{array}{c}
CH_3\!-\!C\!=\!CH\!-\!CH_3 \\
| \\
CH_3
\end{array}
$$

(c) A methyl group CH_3 must be bonded to the ring. If there exist three CH_2 groups then the other carbon must be only bonded to one hydrogen. This must occur at the carbon where a side group is attached to a ring. Thus, the answer must be:

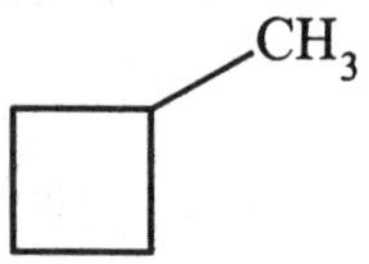

$$
\square\!-\!CH_3
$$

(d) An isopropyl group is

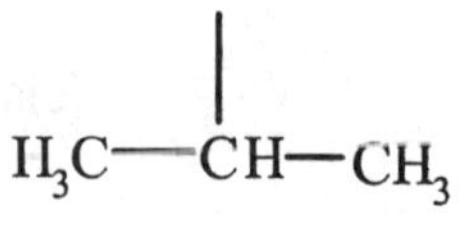

$$
\begin{array}{c}
| \\
H_3C\!-\!CH\!-\!CH_3
\end{array}
$$

The other two carbons must be attached to the isopropyl group and therefore, the double bond must be between these two carbons. Thus, the final correct structure is:

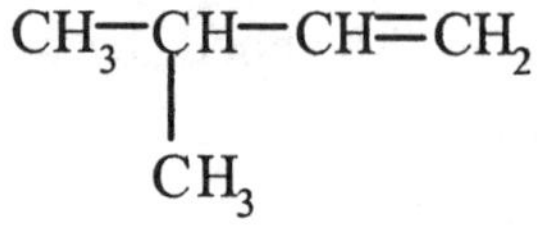

$$
\begin{array}{c}
CH_3\!-\!CH\!-\!CH\!=\!CH_2 \\
| \\
CH_3
\end{array}
$$

12.13 (a) Because the first structure has one Cl and the second has two Cl, they must be different structures that are not isomers.

(b) Because the first structure has four carbon atoms and the second has five carbon atoms, they must be different structures that are not isomers.

(c) Because both structures have four carbons and eight hydrogens and yet cannot be superimposed, they must be constitutional isomers.

12.15 (a) 2-methyl-1-butene: Butene means that the largest chain has four carbons in it and a double bond. The 1 before the butene means that the first carbon in the chain is part of the double bond. The 2-methyl means that there is also a methyl group on the second carbon. Thus the correct structure is:

$$CH_2{=}C{-}CH_2{-}CH_3$$
$$|$$
$$CH_3$$

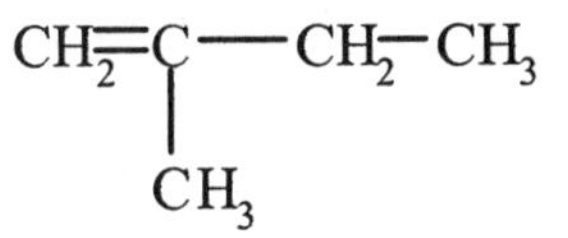

(b) 3-ethyl-2-pentene: Pentene means that the largest chain has five carbons in it and a double bond. The 2 before the pentene means that the first carbon in the chain is part of the double bond. The 3-ethyl means that there is also an ethyl group on the third carbon. Thus the correct structure is:

$$CH_2{-}CH_3$$
$$|$$
$$CH_3{-}CH{=}C{-}CH_2{-}CH_3$$

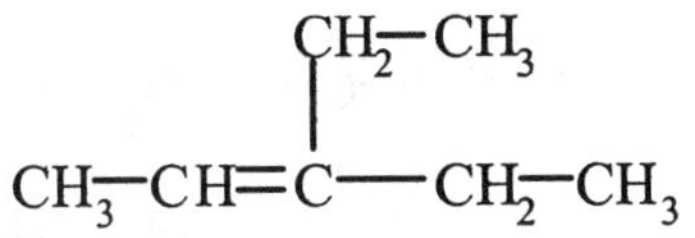

(c) 4-isopropyl-2,6-dimethyl-2-heptene: Heptene means that the largest chain has seven carbons in it and a double bond. The 2 before the heptene means that the second carbon in the chain is part of the double bond. The 4-isopropyl means that there is also an isopropyl group on the fourth carbon. 2,6-dimethyl means that there are two methyl groups attached to the main chain at the second and sixth carbons. Thus the correct structure is:

$$CH_3 \qquad\qquad CH_3$$
$$| \qquad\qquad\quad |$$
$$CH_3{-}C{=}CH{-}CH{-}CH_2{-}CH{-}CH_3$$
$$|$$
$$CH_3{-}CH{-}CH_3$$

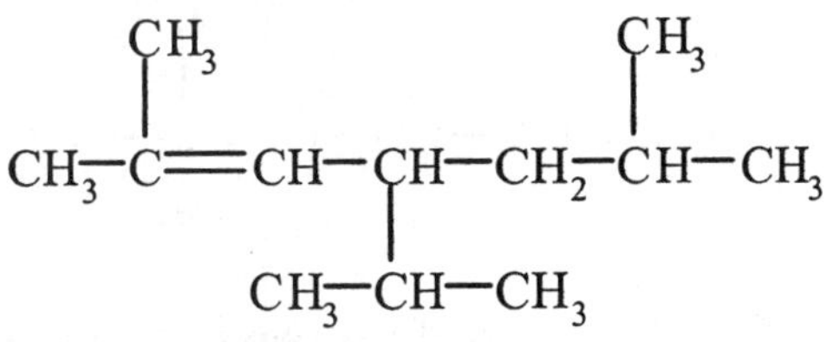

(d) 1,3-dimethylsyclohexene: Cyclohexene means that there is a six-membered ring with a double bond in it. This double bond defines the "1" carbon. The 1,3-dimethyl means that there are two methyl groups attached to the ring at the first and third carbons. Thus the correct structure is:

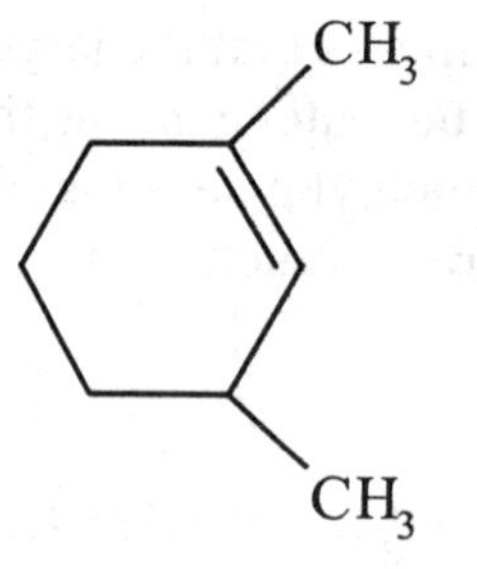

(e) 3-*t*-butyl-2,4-dimethyl-1-pentene: Pentene means that the largest chain has five carbons in it and a double bond. The 1 before the pentene means that the first carbon in the chain is part of the double bond. The 3-*t*-butyl means that there is also a tert-butyl group on the third carbon. 2,4-dimethyl means that there are two methyl groups attached to the main chain at the second and fourth carbons. Thus the correct structure is:

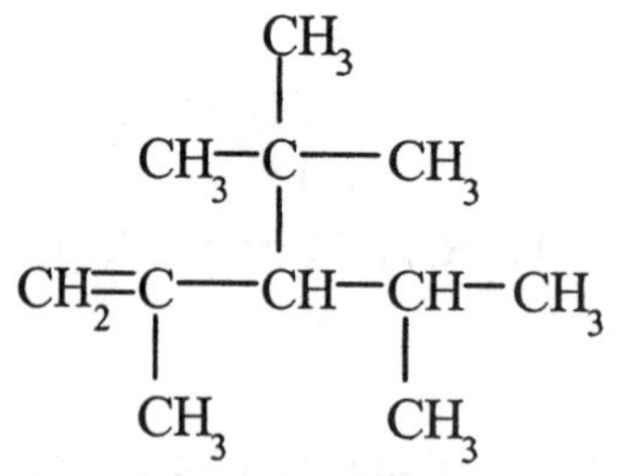

(f) 5-methyl-1,4-hexadiene: Hexadiene means that the largest chain has six carbons in it and two double bonds. The 1,4 before the hexadiene means that the first carbon and the fourth carbon in the chain are part of the double bond. The 5-methyl means that there is also a methyl group on the fifth carbon. Thus the correct structure is:

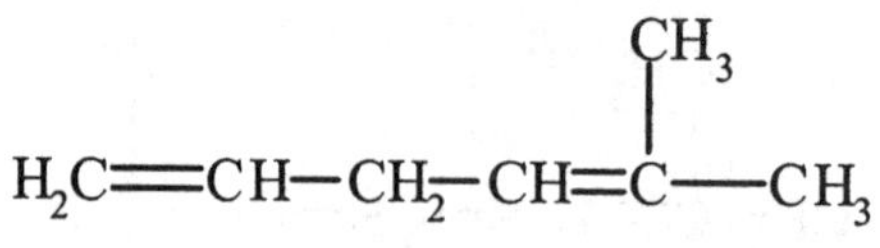

12.17 (a) The first carbon of the double bond has two methyl groups attached to it, therefore this molecule cannot have cis-trans isomers.

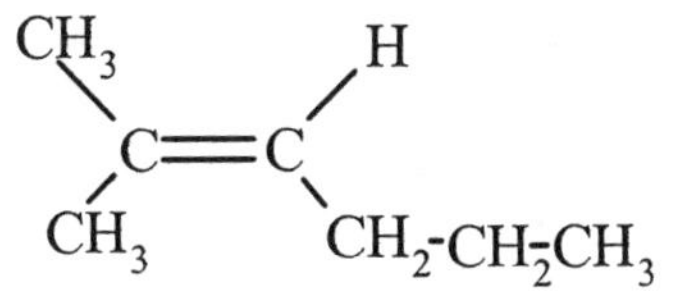

$$CH_3\text{—}C(CH_3)=C(H)\text{—}CH_2\text{-}CH_2CH_3$$

(b) 3-methyl-2-hexene is

$$CH_3,\ H \diagdown C=C \diagup CH_2\text{—}CH_2\text{—}CH_3,\ CH_3 \qquad\qquad H,\ CH_3 \diagdown C=C \diagup CH_2\text{—}CH_2\text{—}CH_3,\ CH_3$$

As both carbons have two different groups, this can form cis-trans isomers as shown above.

(c) 4-methyl-2-hexene is

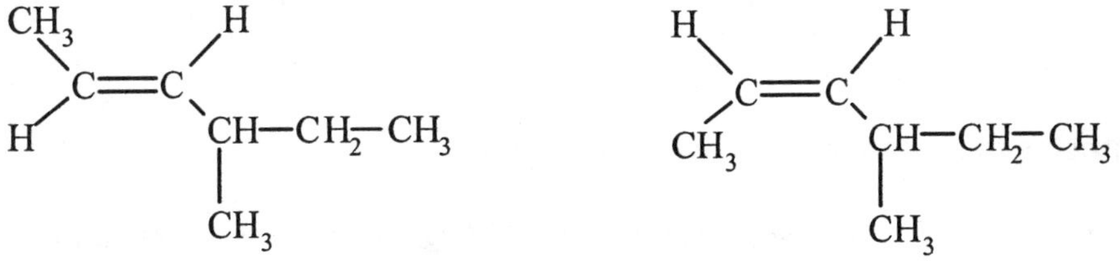

$$CH_3,\ H \diagdown C=C \diagup CH\text{—}CH_2\text{—}CH_3\ (CH_3) \qquad\qquad H,\ H \diagdown C=C \diagup CH\text{—}CH_2\text{—}CH_3\ (CH_3),\ CH_3$$

As both carbons have two different groups, this can form cis-trans isomers as shown above.

(d) 2-methyl-1-hexene is

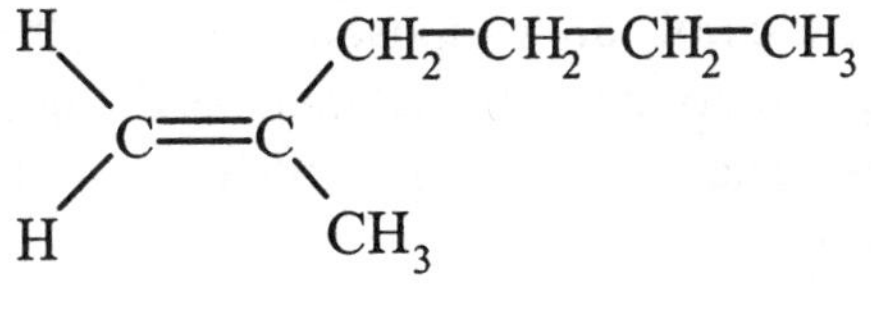

$$H,\ H \diagdown C=C \diagup CH_2\text{—}CH_2\text{—}CH_2\text{—}CH_3,\ CH_3$$

The first carbon of the double bond has two hydrogens attached to it, therefore this molecule can not have cis-trans isomers.

(e) 1,2-dimethylcyclopentene is

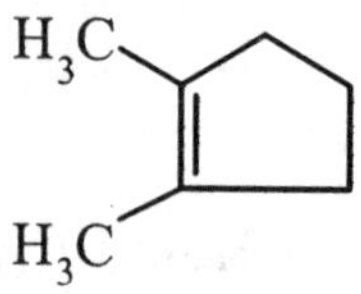

The two carbons that have methyl groups do not have another bond outside the ring; therefore this molecule cannot have cis-trans isomers.

(f) 1,2-dimethylcyclopentane

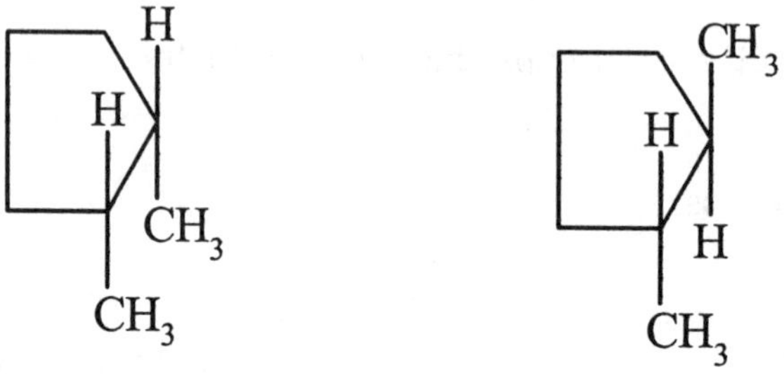

This can form cis-trans isomers as above.

12.19 (a) Both compounds have a formula of $C_4H_6Cl_2$, therefore they are isomers or the same compound. Because there is no free rotation about a double bond, the two compounds cannot be superimposed on each other, therefore they are not the same compound. Inspection shows that they are cis-trans isomers of each other.

(b) If the first compound is rotated about the C=C double bond, it will become the second compound, therefore these two are the same compound.

(c) These two compounds are not superimposed on each other, they are not the same compound. They are also not cis-trans isomers as the compound on the right cannot have cis-trans isomers. They do have the same chemical formula and therefore, they must be constitutional isomers.

(d) The first compound has four carbons while the second has five carbons, therefore these two compounds must be different and not isomers.

12.21 (a) Bromine adds to double bonds at both carbons, therefore the result of the reaction is in both cis and trans form:

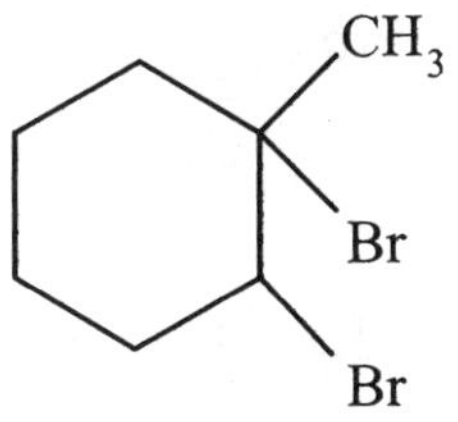

(b) Water adds to the double bond. Note that Markovnikov's rule means that hydrogen adds to the carbon with more H's in the reactant structure.

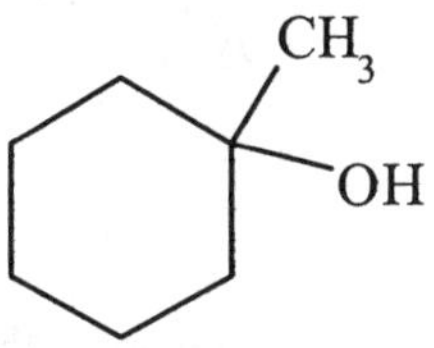

(c) Hydrogen adds to the double bond in the presence of nickel to form:

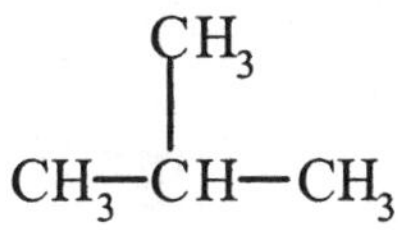

$$CH_3-CH-CH_3$$
$$\overset{\displaystyle CH_3}{|}$$

(d) HCl adds to a double bond. Note that Markovnikov's rule means that hydrogen adds to the primary carbon (the one with more H's in the reactant structure).

$$CH_3-\overset{\displaystyle Cl}{\underset{\displaystyle CH_3}{\overset{|}{\underset{|}{C}}}}-CH_3$$

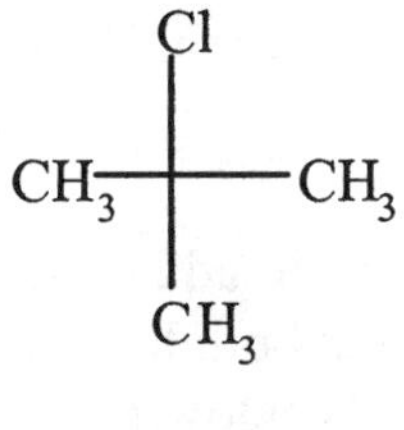

(e) Chlorine adds to the double bond to form:

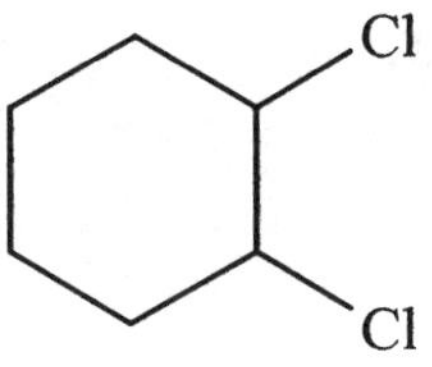

(f) HCl adds to a double bond to form:

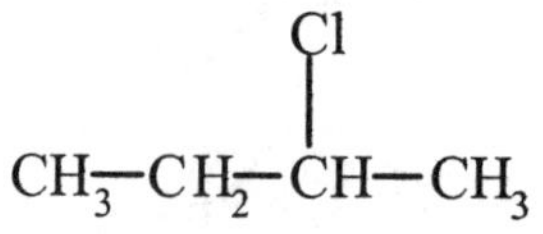

$$CH_3-CH_2-\overset{\overset{\displaystyle Cl}{|}}{CH}-CH_3$$

(g) HBr adds to double bond to form:

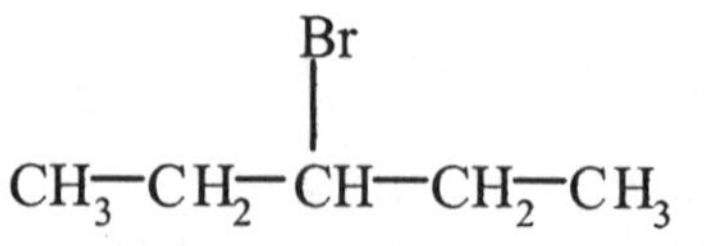

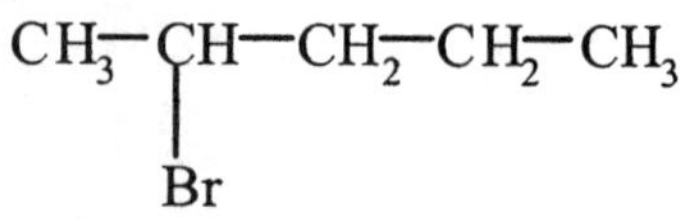

$$CH_3-CH_2-\overset{\overset{\displaystyle Br}{|}}{CH}-CH_2-CH_3 \quad \& \quad CH_3-\underset{\underset{\displaystyle Br}{|}}{CH}-CH_2-CH_2-CH_3$$

Both compounds are formed as the two carbons that form the double bond are equivalent.

(h) Br$_2$ adds to the double bond to form:

$$CH_3-\overset{\overset{\displaystyle Br}{|}}{CH}-\overset{\overset{\displaystyle Br}{|}}{CH}-CH_2-CH_3$$

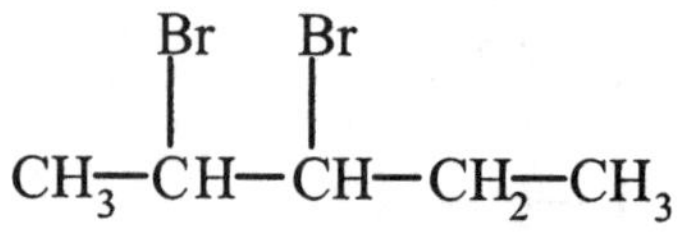

(i) Cl$_2$ also adds to the double bond to form:

$$CH_3-\overset{\overset{\displaystyle Cl}{|}}{CH}-\overset{\overset{\displaystyle Cl}{|}}{CH}-CH_2-CH_3$$

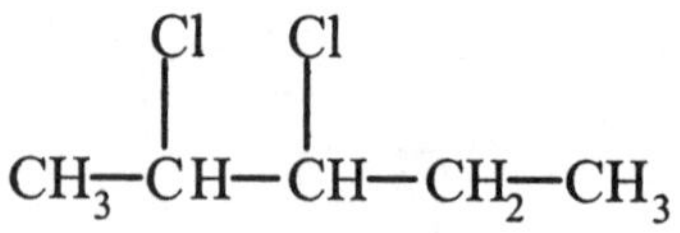

(j) In the presence of an acid, water will add to the double bond. Because the two carbons that form the double bond both have a single hydrogen, they are equivalent and therefore, the two compounds below form:

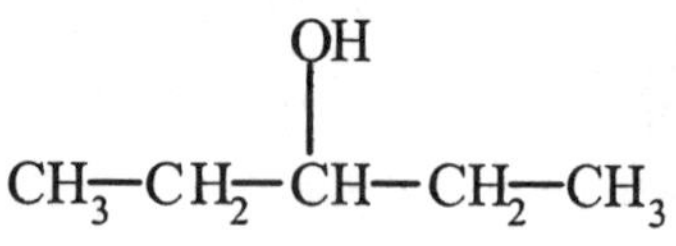

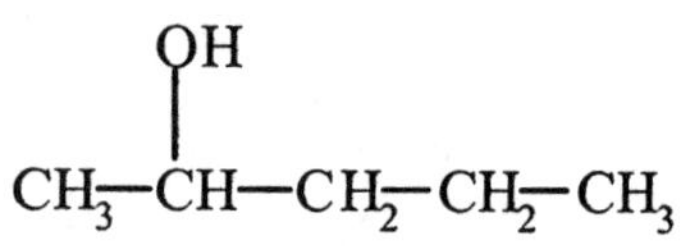

$$CH_3-CH_2-\overset{\overset{\displaystyle OH}{|}}{CH}-CH_2-CH_3 \quad \& \quad CH_3-\overset{\overset{\displaystyle OH}{|}}{CH}-CH_2-CH_2-CH_3$$

(k) Hydrogen adds to the double bond in the presence of nickel to form:

$$CH_3-CH_2-CH_2-CH_2-CH_3$$

12.23   An alkane is not very reactive, while an alkene will undergo addition reactions under the proper conditions. One possibility of this is the addition of bromine (Br_2) to the double bond. The completion of this reaction can be verified by the loss of the red color of Br_2 when the reaction proceeds. Because the red color of bromine does not disappear in this process, there must be no reaction with Br_2. Therefore, the compound must not be an alkene, it must be an alkane, either hexane or 1-methylcyclopentane.

12.25   During a polymerization, the double bond of an alkene is opened to form two more single bonds on the carbons that grow into polymer chains by adding to other double bonds.

$$n \quad H_2C = \underset{\underset{CN}{|}}{CH} \quad \xrightarrow{catalyst} \quad \left(CH_2 - \underset{\underset{CN}{|}}{CH} \right)_n$$

12.27   For the combustion of an alkene, all of the carbon of the alkene goes to carbon dioxide, while all of the hydrogen of the alkene goes to water. Therefore, the balanced equation must be

$$C_nH_{2n} + 3n/2 \; O_2 \rightarrow n \; CO_2 + n \; H_2O$$

For 3-hexene, $n = 6$, therefore the equation is

$$C_6H_{12} + 9 \; O_2 \rightarrow 6 \; CO_2 + 6 \; H_2O$$

12.29   An alkane is not very reactive, while an alkene will undergo reactions under the proper conditions. One possibility of this is the reaction of chromate to the double bond. The completion of this reaction can be verified by the loss of the orange color of chromate when the reaction proceeds. Because the orange color does not disappear in this process, the compound must not be an alkene. It therefore must be an alkane, either pentane or cyclopentane.

12.31   Pentyne means that there are five carbons in the main chain and a triple bond. The 1 before the pentyne signifies that the triple bond is on the first carbon. The 3,4 dimethyl means that there are two methyl groups connected to the main chain at the 3 and 4 carbons. Thus,

$$CH \equiv C - \underset{\underset{CH_3}{|}}{CH} - \underset{\underset{CH_3}{|}}{CH} - CH_3$$

12.33   Aromatic compounds have three double bonds alternating with three single bonds in a 6-membered ring. Aromatic compounds are very stable and do not undergo addition reactions easily. Substitution reactions (replace a hydrogen with another functional group) occur without loss of aromatic character.

12.35   (a) If the first compound is rotated about the center of the aromatic ring, it will become the second compound, therefore these two are the same compound.

(b) Both compounds have a formula of C_7H_7Cl, therefore they are isomers or the same compound. The two compounds cannot be superimposed on each other, therefore they are not the same compound. Inspection shows that they are constitutional isomers of each other.

(c) If the first compound is flipped about the axis defined by the carbon between the CH_3 and Cl groups, it will become the second compound, therefore these two are the same compound.

12.37   (a) A benzene ring with a CH_3 group on it is toluene. This CH_3 group defines the "1" position in the ring. Therefore the chloro group is in the 3 position or "meta" to the methyl group. Thus, 3-chlorotoluene or *m*-chlorotoluene is the correct IUPAC name.

(b) A benzene ring with a CH_3 group on it is toluene. This CH_3 group defines the "1"position in the ring. Therefore the bromo group is in the 4 position or "para" to the methyl group. Thus, 4-bromotoluene or *p*-bromotoluene is the correct IUPAC name.

(c) A benzene ring with a $CH_2C_6H_5$ group is the benzyl group. The chloro group is three carbons away from it or in the "meta" position. Thus, 1-benzyl-3-chlorobenzene or *m*-benzylchlorobenzene is the correct IUPAC name.

(d) A benzene ring with a OH group on it is phenol. This OH group defines the "1" position in the ring. Therefore the bromo group is in the 2 position or "ortho" to the OH group. Thus, 2-bromophenol or *o*-bromophenol is the correct IUPAC name.

(e) A benzene ring with a NH_2 group on it is aniline. This NH_2 group defines the "1" position in the ring. Therefore the ethyl group is in the 4 position or "para" to the NH_2 group. Thus, 4-ethylaniline or *p*-ethylaniline is the correct IUPAC name.

(f) The C_6H_5 group that is attached to the cyclohexane ring is a phenyl group, thus the structure is cis-1-chloro-2-phenylcyclohexane.

12.39   Aromatic compounds undergo limited substitution reactions including halogenation, nitration, sulfonation, and alkylation under the proper conditions.

(a) Aromatics do not react with bases, therefore there is no reaction.

(b) Substitution of bromine only occurs in the presence of a metal halide catalyst. This catalyst is not present, so no reaction occurs.

(c) Sulfonation occurs to form:

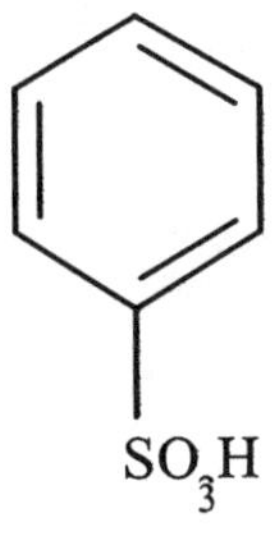

(d) Nitration occurs to form:

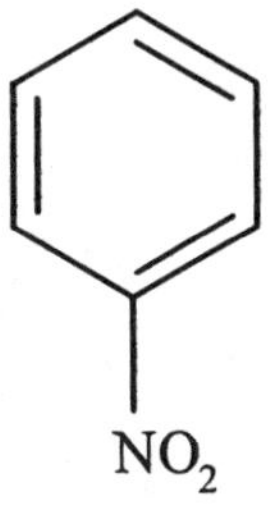

(e) Combustion with O_2 occurs to form:

$$2\ C_6H_6 + 15\ O_2 \rightarrow 12\ CO_2 + 6\ H_2O$$

12.41 The structural formula of benzene is C_6H_6. Therefore, the compound contains another atom each of C, H, and Br. The structures where this is possible are:

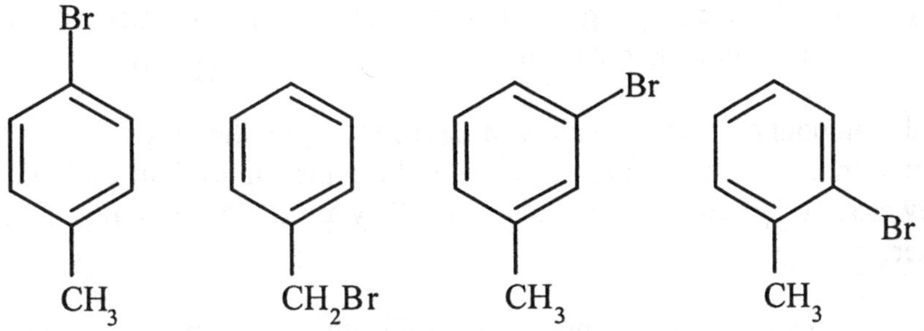

12.43 For a compound to exhibit cis-trans isomers, it must be a cycloalkane or cycloalkene with two single-bonded carbons that each has two different groups attached, or an alkene where both carbons of the double bond have two different substituents attached.

(a) 2-hexene is an alkene that can form cis-trans isomers:

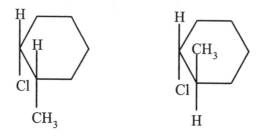

(b) Each of the carbons in a benzene ring has a single substituent, and therefore, an aromatic ring cannot exhibit cis-trans isomerism.

(c) In this cyclohexane, two carbons have different substituents to give the isomers:

(d) The 1 and 2 carbons of a cycloalkene are part of the double bond and can have only one substituent. Therefore, this structure cannot exhibit cis-trans isomerism.

(e) Alkynes only have a single group connected to the carbons that are part of the triple bond (other than the other carbon), and therefore cannot exhibit cis-trans isomerism.

(f) Alkynes only have a single group connected to the carbons that are part of the triple bond (other than the other carbon), and therefore cannot exhibit cis-trans isomerism.

12.45   (a) Both compounds have a formula of C_8H_8; therefore they are isomers or the same compound. The two compounds cannot be superimposed on each other; therefore they are not the same compound. Thus, they are constitutional isomers of each other.

(b) Both compounds have a formula of $C_7H_6Cl_2$; therefore they are isomers or the same compound. The two compounds cannot be superimposed on each other; therefore they are not the same compound. Thus, they are constitutional isomers of each other.

(c) The formula for the first compound is $C_7H_6Cl_2$ while that of the second is $C_7H_{11}Cl_2$; therefore they must be different compounds that are not isomers.

(d) Both compounds have a formula of $C_7H_6Cl_2$; therefore they are isomers or the same compound. The first compound is aromatic while the other is not; therefore they are not the same compound. Thus, they are constitutional isomers of each other.

(e) Both compounds have a formula of C_4H_8; therefore they are isomers or the same compound. The two compounds cannot be superimposed on each other; therefore they are not the same compound. Thus, they are constitutional isomers of each other.

(f) The formula for the first compound is C_4H_8 while that of the second is C_4H_6 Therefore, they must be different compounds that are not isomers.

12.47 All three compounds are nonpolar with very similar boiling points and negligible solubility in water.

12.49 (a) All hydrocarbons will burn (undergo combustion in air) to form CO_2 and H_2O, thus all three will react.

(b) Alkanes, alkenes, and aromatics do not react with a base; therefore none of the three will react.

(c) Only the double bond of cyclohexene will react with an acid such as HCl. The alkane and aromatic structure are too stable to react under these conditions.

(d) Only the double bond of cyclohexene will react with water in the presence of an acid. The alkane and aromatic structure are too stable to react under these conditions.

(e) Only the double bond of cyclohexene will react with the oxidizing agent $KmnO_4$. The alkane and aromatic structure are too stable to react under these conditions.

12.51 The bromine will react with both double bonds to form:

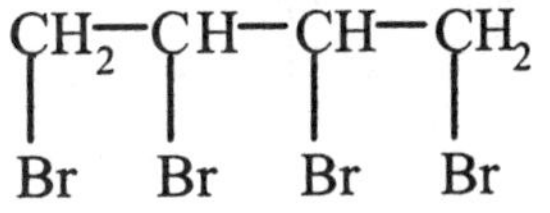

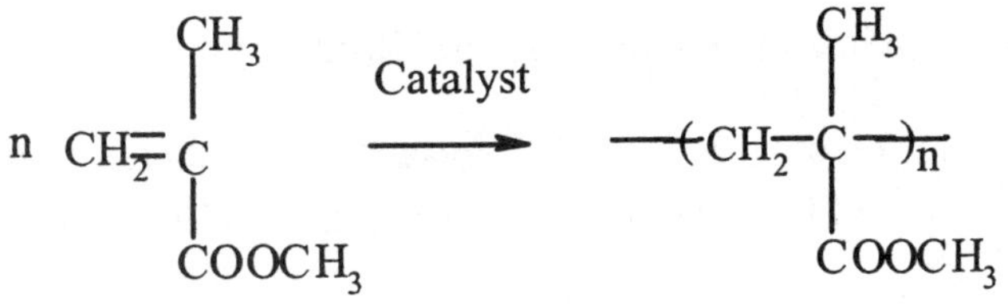

$$\underset{\underset{Br}{|}}{CH_2}-\underset{\underset{Br}{|}}{CH}-\underset{\underset{Br}{|}}{CH}-\underset{\underset{Br}{|}}{CH_2}$$

$$n\ \underset{\underset{COOCH_3}{|}}{\overset{\overset{CH_3}{|}}{CH_2\!\!=\!\!C}}\ \xrightarrow{\text{Catalyst}}\ \underset{\underset{COOCH_3}{|}}{\overset{\overset{CH_3}{|}}{-\!\!(CH_2\!\!-\!\!C\!\!-\!\!)_n}}$$

12.53 (a) Counting hydrogens and carbons gives C_6H_8.

(b) Counting hydrogens and carbons gives C_9H_{12}.

(c) Counting hydrogens and carbons gives C_5H_8.

(d) Counting hydrogens and carbons gives $C_6H_7BrClNO_2$.

12.55   An alkyne such as 1-hexyne will undergo an addition reaction to form a dibromoalkene. This alkene can then undergo a further addition reaction to form an alkane, which uses up more bromine. Thus the reaction of the alkyne with Br_2 will utilize twice as much Br_2 as the alkene. This leads to the conclusion that A is 1-hexene, while B is 1-hexyne.

12.57   (a) The double bonds will add the Br_2 to all four carbons to form:

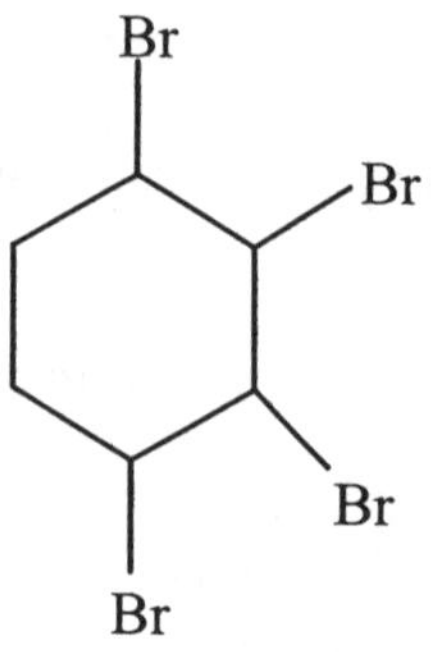

(b) Benzene does not react with Br_2 in the absence of a catalyst; no reaction.

(c) Benzene will undergo a substitution reaction with Br_2 in the presence of $FeBr_2$ to form:

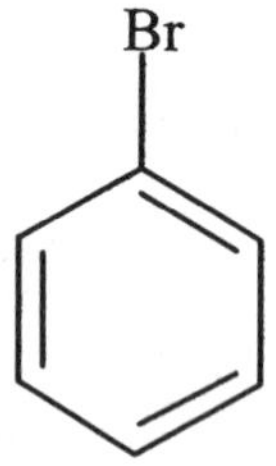

(d) Alkenes do not react with bases; no reaction.

(e) Alkanes do not oxidize in the presence of $KMnO_4$; no reaction.

(f) Benzene will not react with Cl_2 in the presence of heat and light, however alkanes will. Thus the CH_3 group will replace a hydrogen with a chlorine to form:

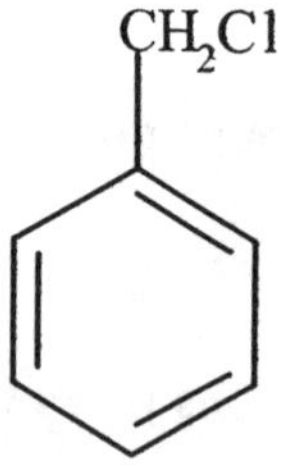

(g) Benzene will undergo a halogenation reaction in the presence of $FeCl_2$. Toluene will replace a hydrogen with a chlorine to form the following compounds:

(h) All hydrocarbons will undergo combustion to form CO_2 and H_2O.

(i) All hydrocarbons will undergo combustion to form CO_2 and H_2O.

(j) Benzene is too stable to react with HCl; no reaction.

(k) Cyclohexene will add HCl across the double bond to form:

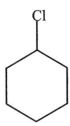

12.59 The change in color signifies a reaction with the dichromate. An alkane is fairly stable and will not react with dichromate while the double bond of an alkene will. Therefore, the unknown must be an alkene, 1-hexene.

12.61 Mary Jones is correct. Addition of water to propene will proceed according to Markovnikov's rule to produce 2-propanol as the major product, not 1-propanol.

12.63 Perform the bromine test in a quantitative manner. Weigh out an amount of the shipped compound. Calculate the number of moles assuming the compound is geraniol. Determine the amount of bromine that reacts with the compound. Calculate the number of moles of bromine used. If the compound were geraniol, the number of moles of bromine would be twice the moles of the compound. If more than twice the moles of bromine are used, the shipped compound is most likely myrcene. This can be verified by performing the calculation assuming the compound is myrcene. Myrcene would react with three times the number of moles of bromine.

12.65   Unsaturated fats contain carbon-carbon double bonds, saturated fats contain only single bonds.

12.67   The reaction rate increases with increasing stability of the carbocation formed when H^+ adds to the double bond of each alkene. More stable carbocations are formed faster than less stable carbocations. 2-Methylpropene forms a tertiary carbocation, propene forms a secondary carbocation, and ethene forms a primary carbocation. The order of carbocation stability is $3° > 2° > 1°$ and this results in the observed order of reactivity of alkenes.

12.69   The HCl can add across both double bonds to give:

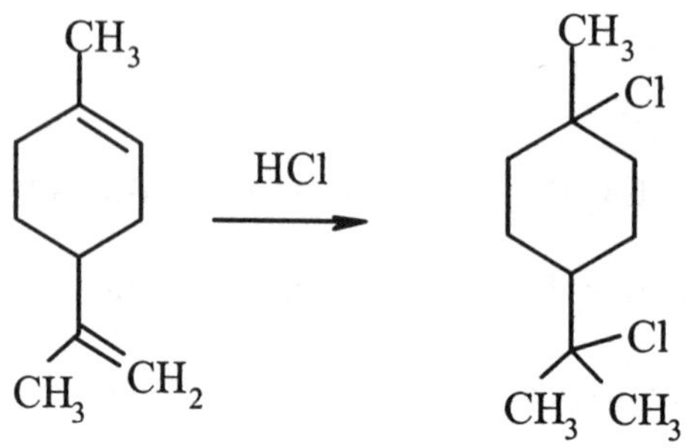

12.71   The methyl groups attached to the benzene ring will undergo oxidation under moderately strong conditions, such as hot acidic $K_2Cr_2O_7$.

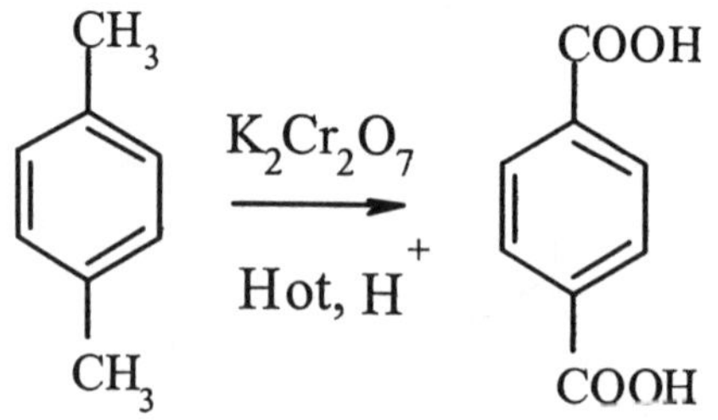

12.73   Limonene contains carbon-carbon double bonds that can react with Br_2 by addition, while the vanillin does not have these double bonds. Thus, limonene reacts with Br_2, which alters the color of the mixture from red to clear, while the vanillin does not react and the red color of the Br_2 remains.

12.75   The unknown is *p*-dimethylbenzene, (A) below, as the substitution of the Br onto the benzene ring will result in the same compound, 2-bromo-1,4-dimethylbenzene, (B) below.

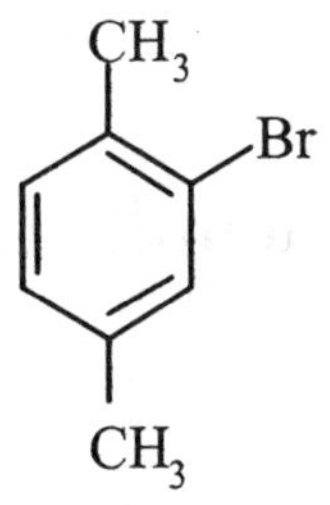

m-dimethylbenzene would yield three monobromo products, while *o*-dimethylbenzene would yield two monobromo products.

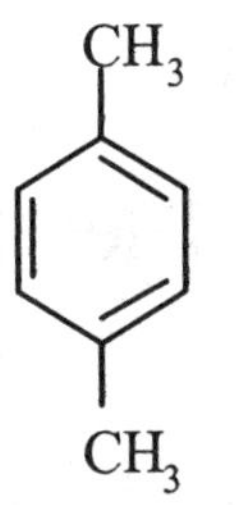

12.77 *m*-methylnitrobenzene is produced by first nitrating benzene followed by alkylation:

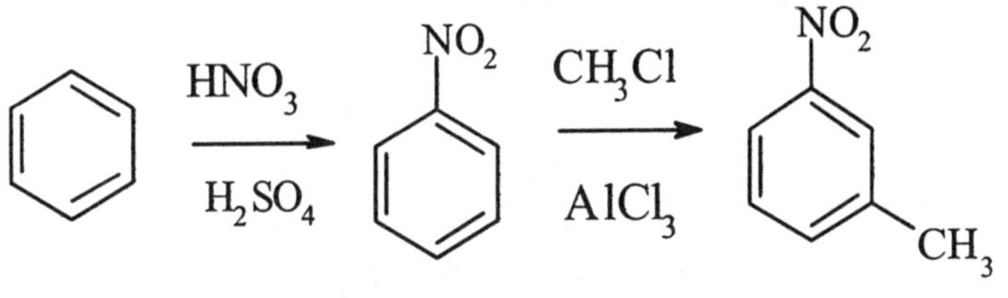

Alkylation followed by nitration would yield a mixture of the *o*- and *p*-isomers, not the *m*-isomer.

12.79 The C–O bond is too strong to break in the absence of acid. When acid is present, the oxygen is protonated to yield C–O$^+$H$_2$, where the + charge on the oxygen weakens the C–O bond and H$_2$O is expelled.

12.81 The oxidation state of carbon increases as the number of bonds to hydrogen decreases and/or the number of bonds to oxygen increases. As each carbon atom of ethene is bonded to fewer hydrogens than each carbon atom of ethane, the oxidation state of the carbons in ethene are higher than those of ethane.

Chapter 13

Alcohols, Phenols, Ethers, and Their Sulfur Analogues

13.1 An alcohol is an organic compound that has an –OH bonded to a single non-aromatic carbon. A phenol has an –OH bonded to a carbon that is part of an aromatic ring. An ether is a compound with an oxygen that is bonded to two carbons.

13.3 (a) A tert-butyl group is $-C(CH_3)_3 = C_4H_9$. Therefore, CH_3O remains to be bonded to the tert-butyl group. Moreover, as the compound is an alcohol, the CH_3O must be bonded as CH_2–OH. Therefore, the compound must be:

$$CH_3-\underset{\underset{\displaystyle CH_3}{|}}{\overset{\overset{\displaystyle CH_3}{|}}{C}}-CH_2-OH$$

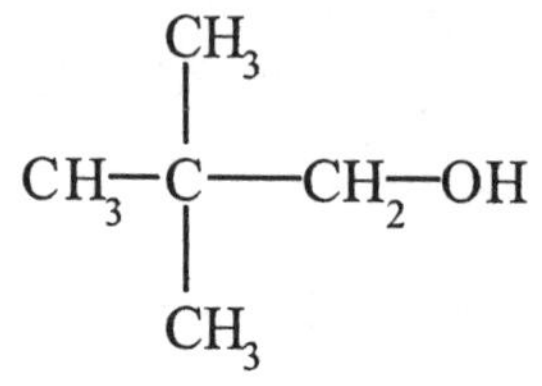

(b) One methyl group = CH_3, thus the rest of the compound consists of C_3H_7O. As the compound is an alcohol, one of the hydrogens must be bonded to the oxygen. Therefore, the compound must be:

$$CH_3-CH_2-CH_2-CH_2-OH$$

13.5 Counting carbons, hydrogens, and oxygens:

(a) $C_6H_{14}O = C_nH_{2n+2}O$ (b) $C_6H_{12}O = C_nH_{2n}O$

13.7 Method for naming alcohols:

(i) Find longest chain (or ring) that contains the –OH. Form base name, by adding "ol" to the end of the base alkane.

(ii) Put a number before the base name to denote to which carbon the –OH is attached.

(iii) Determine side groups that are attached to the base chain.

(iv) Put names of these groups in alphabetical order at beginning of name. Include numbers to denote to which carbon(s) the groups are attached.

(v)   When classifying the alcohol as primary, secondary or tertiary, count the number of carbons that are attached to the carbon on which the –OH is bonded. 1°, 2°, and 3° carbons are directly bonded to one, two, and three other carbon atoms, respectively.

(a) (i) 4 carbons = butane → butanol.

(ii) –OH is connected to the first carbon, therefore, 1-butanol.

(iii) There is also a CH_3 (= methyl) group connected to the C-2.

(iv) <u>2-methyl-1-butanol.</u>

(v) That carbon is attached to one other carbon atom, therefore it is a <u>primary alcohol</u>.

(b) (i) 4 carbons = butane → butanol.

(ii) –OH is connected to the second carbon, therefore, 2-butanol.

(iii) There is also a CH_3 (= methyl) group connected to the C-3.

(iv) <u>3-methyl-2-butanol.</u>

(v) That carbon is attached to two other carbon atoms, therefore it is a <u>secondary alcohol</u>.

(c) (i) 4 carbons = butane → butanol.

(ii) –OH is connected to the first carbon, therefore, 1-butanol.

(iii) There is also a CH_3 (= methyl) group connected to the C-2.

(iv) <u>2-methyl-1-butanol.</u>

(v) That carbon is attached to one other carbon atom, therefore it is a <u>primary alcohol</u>.

(d) (i) 3 carbons = propane → propanol.

(ii) –OH is connected to the second carbon, therefore, 2-propanol.

(iii) There is also a CH_3 (= methyl) group connected to the C-2.

(iv) <u>2-methyl-2-propanol.</u>

(v) That carbon is attached to three other carbon atoms, therefore it is a <u>tertiary alcohol</u>.

13.9   The alcohols have stronger secondary forces than alkanes and thus have higher boiling points than alkanes (hydrogen bonds are stronger than London dispersion forces). Therefore, hexane will have the lowest boiling point in this list. Within the group of alcohols, the linear alcohol molecules will interact with other linear alcohols more efficiently than branched alcohols, thus 2-methyl-2-butanol will have the lowest boiling point of the alcohols. Of the remaining two, 1,4-butanediol will have stronger secondary interactions than 1-pentanol because it can form two hydrogen bonds per molecule while 1-pentanol can only form one. Thus:

1,4-butanediol B.P. > 1-pentanol B.P. > 2-methyl-2-butanol B.P. > hexane B.P.

13.11 Ethanol is not sufficiently acidic to alter the color of blue litmus paper, therefore there will be no change.

13.13 Strong acids will slightly protonate an alcohol. Thus the reaction will be:

$$CH_3CH_2CH_2CH_2-OH \ + \ HCl \ \rightleftharpoons \ CH_3CH_2CH_2CH_2-OH_2^+ \ + \ Cl^-$$

13.15 (a) In the absence of heat, alcohols will accept a proton from very strong acids, thus an acid-base reaction will occur:

$$CH_3CH_2CH_2CH_2-OH \ + \ H_2SO_4 \ \rightleftharpoons \ CH_3CH_2CH_2CH_2-OH_2^+ \ + \ HSO_4^-$$

(b) Alcohols do not decompose in the presence of heat, thus no reaction occurs.

(c) In the presence of heat and a strong acid, dehydration, or the loss of water to form a double bond, will occur:

$$CH_2{=}CH-CH_2-CH_3$$

13.17 A primary alcohol will undergo two consecutive oxidations, first to an aldehyde, then the aldehyde will be oxidized further to form a carboxylic acid. A secondary alcohol will oxidize to a ketone while a tertiary alcohol will not undergo oxidation short of combustion. Therefore:

(a)

(b) No reaction.

13.19   Butene will react with the permanganate at the double bond. 1-butanol will also react with permanganate to oxidize the –OH. Consequently, both of these compounds will react (and therefore de-color) a permanganate solution. Therefore, this test can not distinguish between 1-butene and 1-butanol.

13.21   In naming phenols, the position of the –OH on the aromatic ring defines the C-1 position on the ring. Additional substituents on the ring should be included in alphabetical order in front of phenol. Thus, this compound is named 2-isopropyl-5-methylphenol.

13.23   When phenol acts as an acid it donates the hydroxyl hydrogen:

$$p\text{-}CH_3\text{-}C_6H_5\text{—}OH \quad + \quad H_2O \quad \rightleftharpoons \quad p\text{-}CH_3\text{-}C_6H_5\text{—}O^- \quad + \quad H_3O^+$$

13.25   (a) Phenols will burn, i.e., undergo combustion in the presence of oxygen as:

$$C_6H_6 + 7\,O_2 \rightarrow 6\,CO_2 + 3\,H_2O$$

(b) Phenols are acids and therefore will undergo an acid-base reaction in the presence of a base:

$$C_6H_5\text{–}OH + KOH \rightarrow C_6H_5\text{–}O^- + K^+ + H_2O$$

13.27   (a) 4-ethoxyphenol: Phenol as the base name means that the primary structure is an aromatic ring with an –OH on it. The carbon that the –OH is connected to is labeled as C-1. 4-ethoxy means that there is a –$O(C_2H_5)$ group also connected to the aromatic ring at C-5. Thus:

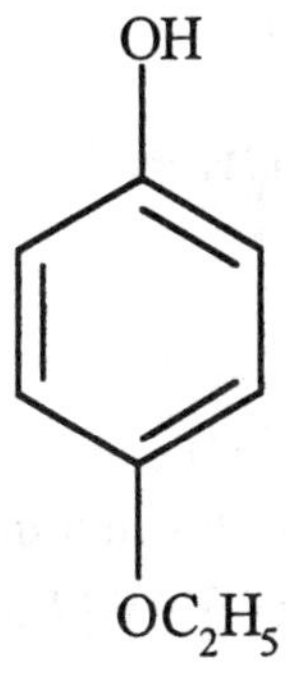

OH

OC$_2$H$_5$

(b) Isobutyl propyl ether: An ether is a compound that connects two organic groups by a single oxygen linkage. The "isobutyl" and "propyl" in the names denotes that structure of the two organic groups that exists on both sides of the oxygen, thus:

$$(CH_3)_2CHCH_2 \!-\!\!-\!\!-\!O\!-\!\!-\!\!-\!CH_2CH_2CH_3$$

(c) 3-isopropoxy-1-butanol: 1-butanol means that the base structure is a four carbon chain (butan-) with an –OH connected to it on C-1. 3-isopropoxy means that there is an isopropyl group that is connected to the main carbon chain by an oxygen (oxy-) linkage at C-3. Thus:

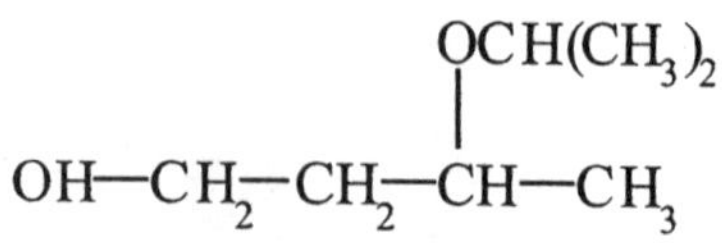

$$\begin{array}{c} OCH(CH_3)_2 \\ | \\ OH\!-\!CH_2\!-\!CH_2\!-\!CH\!-\!CH_3 \end{array}$$

13.29 The boiling point of a substance is controlled by the strength of its secondary forces; a stronger intermolecular force will result in a higher boiling point. Ethanol can form hydrogen bonds among its molecules while the other two compounds cannot; therefore it has the highest boiling point in this list. Between dimethylether and propane, dimethylether will have dipole-dipole interactions between molecules while propane will only have London dispersion forces, thus:

ethanol B.P. > dimethylether B.P. > propane B.P.

13.31 1-methylcyclopentene is formed at both temperatures. Although at 140 °C ether formation is favored for primary alcohols, this alcohol is a tertiary alcohol which will not undergo a reaction to form an ether.

13.33 (a) 2-butanethiol: thiol means that there is an –SH connected to the main chain. Butane means that the main carbon chain has four carbons and the 2- denotes that the –SH is connected to the chain at C-2. Thus:

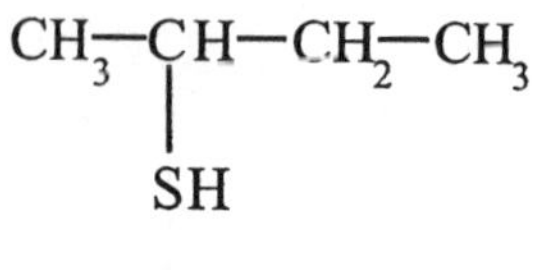

$$\begin{array}{c} CH_3\!-\!CH\!-\!CH_2\!-\!CH_3 \\ | \\ SH \end{array}$$

(b) Dicyclopentyl disulfide: Disulfide means that there exists an –S–S– linkage in this molecule. The dicyclopentyl denotes that the organic structures that are connected to the two ends of the –S–S– linkage are both cyclopentane rings, thus:

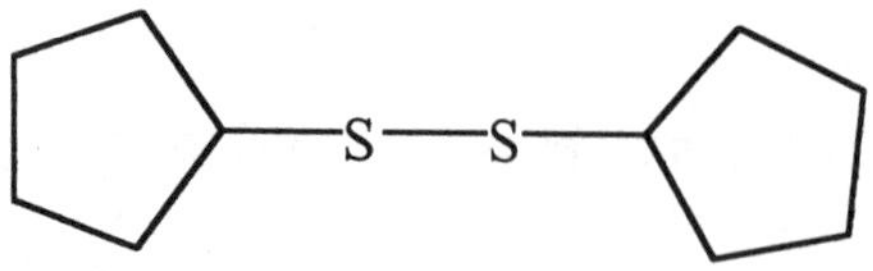

13.35 A, phenol; B, ether; C, alcohol; D, thiol; E, ether.

13.37 An alcohol is a compound that has an –OH bound to an sp^3 carbon, a phenol is a compound that has an –OH group connected to an aromatic carbon, a thiol is a compound with an –SH group, an ether is a compound with an oxygen that is bonded to two carbons, and a disulfide is a compound with a –S–S– in it. Thus:

(a) this compound is a disulfide.

(b) this compound is an alcohol.

(c) this compound is a thiol.

(d) this compound is an ether.

13.39   An ether or an alcohol without a double bond or ring must have a formula that follows $C_nH_{2n+2}O$, while an alcohol or an ether with a double bond or ring will have a formula that follows $C_nH_{2n}O$. Thus this compound can be either an ether or an alcohol, but must have either a double bond or ring (but not both).

13.41   (a) *s*-butyl propyl ether: An ether means that there is an –O– linkage in the main chain with two organic groups on each side. One of them is an *s*-butyl group while the other is propyl, thus:

$$CH_3CH_2CH_2\!\!-\!\!-\!\!O\!\!-\!\!-\!\!CH_2CH(CH_3)_2$$

(b) *m*-propylphenol: Phenol denotes that the compound is an aromatic with an –OH which defines the C-1 position. *m*-propyl denotes that there also exists a propyl group (C_3H_7) on C-3, thus:

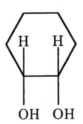

(c) *cis*-1,2-cyclohexanediol: Cyclohexanediol denotes that the molecule is a six-carbon ring with two –OH groups attached. The 1,2 denote that the –OH groups are attached to C-1 and C-2. *Cis* denotes that that the –OH groups are on opposite sides of the ring, thus:

(d) 2-pentanethiol: "Thiol" denotes that there is an –SH on the chain, while 2-pentane denotes that the main carbon chain has five carbons and the –SH is connected at C-2, thus:

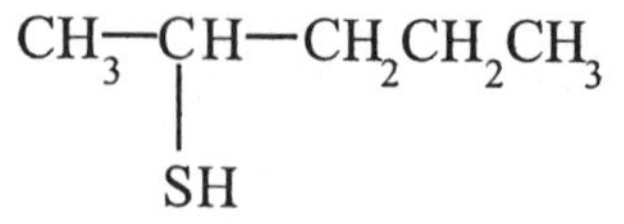

$$CH_3{-}CH{-}CH_2CH_2CH_3$$
$$|$$
$$SH$$

(e) 1-pentene-3-ol: Pentene denotes that the primary structure of the molecule is a five-carbon chain with a double bond. The "1" denotes that the double bond is between C-1 and C-2. The -3-ol at the end of the name denotes that there also exists an –OH on this chain at C-3, thus:

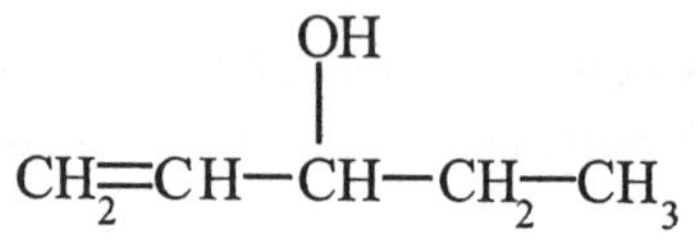

$$OH$$
$$|$$
$$CH_2{=}CH{-}CH{-}CH_2{-}CH_3$$

13.43 A hydrogen bond will form between an O, N, or F atom and a hydrogen that is covalently bonded to an O, N, or F atom. Thus:

(a) will form hydrogen bonds as

$$C_2H_5{-}O\!\!\!\diagdown\!\!\!\diagup\!\!\!-\!\!\!-\!\!\!-\!\!\!-\!\!\!-H{-}O\!\!\!\diagdown$$
$$\qquad\qquad H\qquad\qquad\qquad C_2H_5$$

(b) will form hydrogen bonds as

$$C_3H_7{-}O\!\!\!\diagdown\!-\!-\!-\!-\!-H{-}O\!\!\!\diagdown\qquad\qquad C_3H_7{-}O\!\!\!\diagdown$$
$$\qquad\qquad H\qquad\qquad\qquad H\qquad\quad and\qquad\qquad\qquad H$$
$$\qquad\qquad\qquad\qquad\qquad\qquad\qquad\qquad\qquad\qquad O{-}H$$
$$\qquad\qquad\qquad\qquad\qquad\qquad\qquad\qquad\qquad H$$

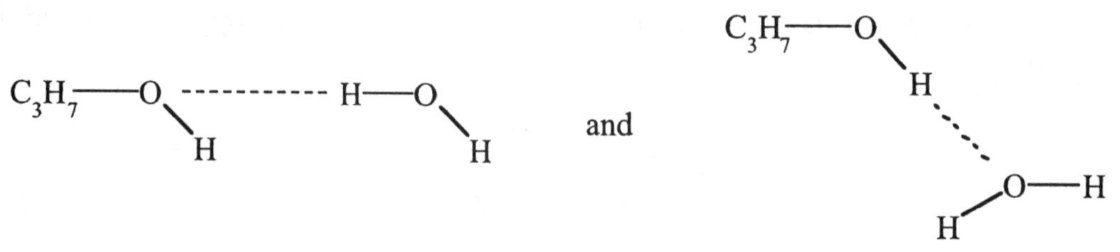

(c) cannot form hydrogen bonds.

(d) will form hydrogen bonds as

$$CH_3-O \cdots\cdots H-O$$
$$\quad\quad\quad\quad CH_3 \quad\quad\quad CH_3$$

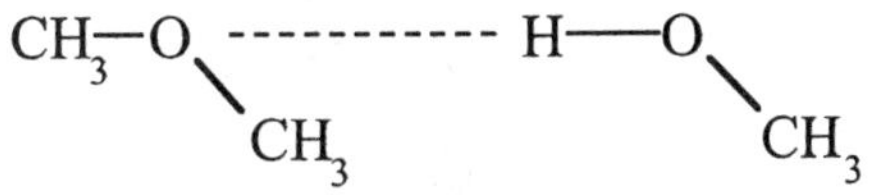

(e) will form hydrogen bonds as

$$C_6H_5-O$$
$$\quad\quad\quad H$$
$$\quad\quad\quad\quad \vdots$$
$$\quad\quad\quad O-H$$
$$\quad H$$

and

$$C_6H_5-O \cdots\cdots H-O$$
$$\quad\quad\quad\quad H \quad\quad\quad\quad H$$

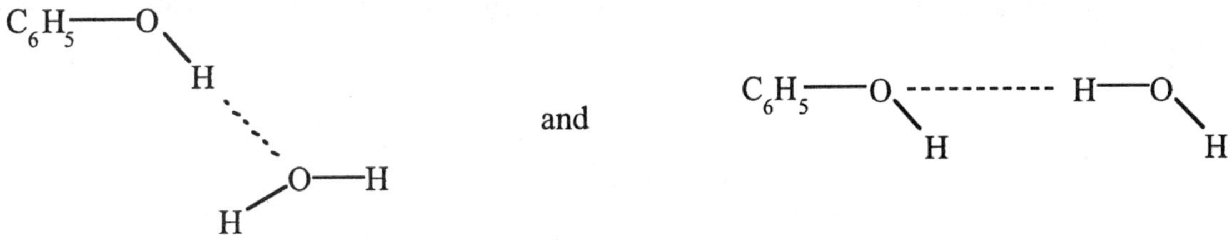

13.45   The molecule that most easily gives up a proton is the most acidic, while the one that is least likely to donate a proton is the least acidic. Given this definition, these compounds arranged in decreasing order of acidity are:

HCl > phenol > cyclohexanethiol > cyclohexanol

13.47   (a) Phenols are moderate acids that will react with a very strong base such as NaOH:

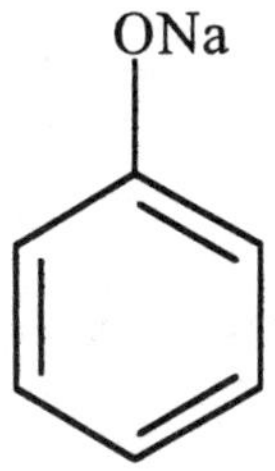

ONa

(b) Ethers are neutral and therefore do not react with bases, thus no reaction.

(c) Thiols are weak acids and will thus undergo an acid-base reaction in the presence of a strong base, thus:

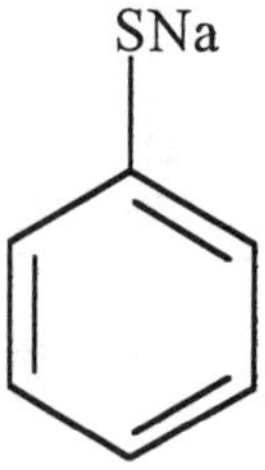

(d) Alcohols are not strong enough acids to react with a base, therefore, no reaction.

(e) Alcohols are very weak acids that will give up a proton to an active metal such as Na:

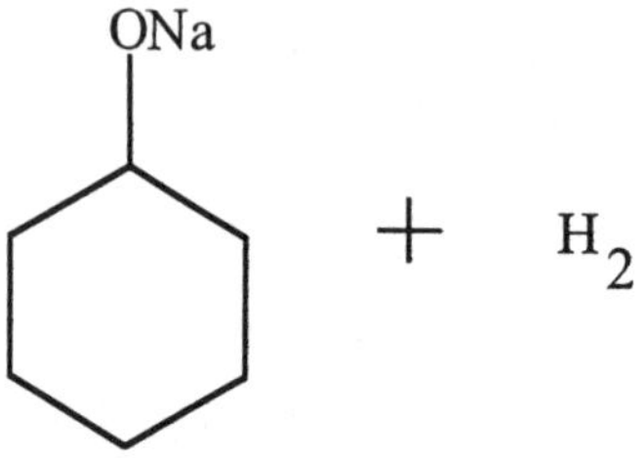

(f) Thiols will react with heavy metals such as Pb to form a salt

$$CH_3CH_2CH_2CH_2{-\!\!-\!\!-}S{-\!\!-\!\!-}Pb{-\!\!-}S{-\!\!-}CH_2CH_2CH_2CH_3$$

(g) The S–H bond in a thiol is weak and therefore a thiol will undergo oxidation to form a disulfide.

$$CH_3CH_2CH_2CH_2{-\!\!-\!\!-}S{-\!\!-}S{-\!\!-}CH_2CH_2CH_2CH_3$$

(h) Disulfides are also easily reduced backed to thiols:

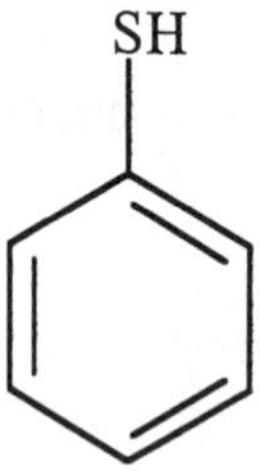

13.49 (a) alcohols are slightly acidic while hydrocarbons are neutral, therefore, 1-propanol.

(b) HCl is a very strong acid while alcohols are merely slightly acidic, therefore, HCl.

(c) Alcohols and water are about the same acidity, therefore the same.

(d) Phenol is more acidic than alcohols due to the stability of the aromatic ring, thus phenol.

(e) Thiols are more acidic than alcohols, therefore, 1-propanethiol.

13.51 Of the compounds listed here, only 1-butanol will react with permanganate. The fact that the unknown does not react with permanganate means that the unknown is not 1-butanol. The unknown could be any of the other compounds, *t*-butyl alcohol, ethyl methyl ether, or pentane, as none of these compounds react with permanganate.

13.53 Of these three compounds, only phenol is acidic enough to turn blue litmus paper red. Therefore, the unknown must be phenol.

13.55 A, C, and E denote the angle about an oxygen in an ether or alcohol linkage. This angle is close to 109.5°. B is the angle about a carbon that is part of an aromatic ring, which equals 120°. Finally, D is the angle about a carbon that is bonded to four atoms which equals 109.5°.

13.57 Test the compound with blue litmus paper. Hexylresorcinol is a phenol and sufficiently acidic that it turns blue litmus paper red. 3-Hexyl-1, 2-cyclohexanediol is an alcohol and is neutral to litmus paper.

13.59 1-Propanol evaporates rapidly from the skin surface, removing heat and creating a cooling effect. 1-Decanol and 2-decanol are much less volatile because of their higher molecular masses, which results in less heat removed from the skin surface.

13.61 Carbocations (positively charged carbon atoms) are intermediates in the reaction. The carbocation formed from 2-propanol is secondary and more stable than the primary carbocation formed from 1-propanol. The more stable carbocation is formed faster and this results in a faster overall rate for the dehydration reaction.

13.63 Ethylene oxide is relatively unstable due to the distortion of its bond angles from the normal tetrahedral bond angle. Reaction to form an acyclic product with tetrahedral bond angles relieves this instability. The six-membered ring of dioxane possesses tetrahedral bond angles and is, therefore, a very stable compound. Opening of the cyclic dioxane structure does not increase stability.

13.65 The dehydration of 2-phosphoglyceric acid involves the loss of water and creation of a carbon-carbon double bond to form:

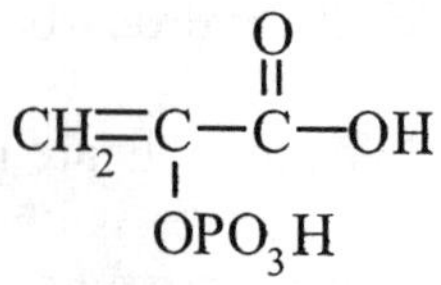

$$\underset{\underset{OPO_3H}{|}}{CH_2} = \underset{}{C} - \overset{\overset{O}{\|}}{C} - OH$$

13.67   The base structure of menthol is a six-member ring that has an –OH bonded to cyclohexanol. The carbon that is bonded to the –OH is C–1, there is a *t*-butyl group at the C–2 position and a methyl group as C–5, thus the correct name is 2-*t*-butyl-5-methylcyclohexanol. 2-*t*-butyl-1-hydroxy-5-methylcyclohexane is also acceptable.

13.69   The acidity is enhanced because the *p*-chlorophenoxide anion is more stable than the phenoxide anion. The *p*-chlorophenoxide anion is more stable than the phenoxide anion because the presence of the Cl on the benzene ring creates a more electron-withdrawing structure.

13.71   Tertiary alcohols do not undergo intermolecular dehydration to form ethers.

13.73 $C_2H_6O + 3O_2 \rightarrow 2CO_2 + 3H_2O$

Chapter 14

Aldehydes and Ketones

14.1   Aldehydes and Ketones have the same C:H ratio as alkene, but also have an oxygen, $C_nH_{2n}O$. Alcohols and ethers have molecular formulas of $C_nH_{2n+2}O$. Thus this molecule is an alcohol or an ether.

14.3   An aldehyde has the structure R-CHO, where R denotes a hydrocarbon structure. This molecule has the formula $C_6H_{12}O$, thus the R group must consist of 5 carbons and 11 hydrogens. The isomers that can be formed from these combinations are:

$$CH_3CH_2CH_2CH_2CH_2\text{—}\overset{\displaystyle O}{\overset{\displaystyle \|}{C}}\text{—}H \qquad CH_3CH_2CH_2\text{—}\overset{\displaystyle CH_3}{\underset{\displaystyle |}{CH}}\text{—}\overset{\displaystyle O}{\overset{\displaystyle \|}{C}}\text{—}H$$

$$CH_3\text{—}\underset{\displaystyle \underset{\displaystyle CH_3}{|}}{\overset{\displaystyle \overset{\displaystyle CH_3}{|}}{CH}}\text{—}CH\text{—}\overset{\displaystyle O}{\overset{\displaystyle \|}{C}}\text{—}H$$

$$CH_3\text{—}\overset{\displaystyle CH_3}{\underset{\displaystyle |}{CH}}\text{—}CH_2\text{—}CH_2\text{—}\overset{\displaystyle O}{\overset{\displaystyle \|}{C}}\text{—}H$$

$$CH_3CH_2\text{—}\overset{\displaystyle CH_3}{\underset{\displaystyle |}{CH}}\text{—}CH_2\text{—}\overset{\displaystyle O}{\overset{\displaystyle \|}{C}}\text{—}H$$

$$(CH_3)_3CCH_2\text{—}\overset{\displaystyle O}{\overset{\displaystyle \|}{C}}\text{—}H$$

$$CH_3CH_2\text{—}\underset{\displaystyle \underset{\displaystyle CH_3}{|}}{\overset{\displaystyle \overset{\displaystyle CH_3}{|}}{C}}\text{—}\overset{\displaystyle O}{\overset{\displaystyle \|}{C}}\text{—}H$$

14.5   A ketone has the general formula R–CO–R', where R and R' are hydrocarbon structures; an aldehyde has the structure R-CHO, where R denotes a hydrocarbon structure. Thus, compound 1 is an aldehyde; compound 2 is not an aldehyde or a ketone, it is an ester (from Table 14.1); compound 3 is a ketone; and compound 4 is not an aldehyde or a ketone, it is an amide (from Table 14.1).

14.7   Atoms a, d, and e are carbons that are bonded to four other atoms, therefore they are sp^3 hybridized. Atoms b and c are atoms that are part of a double bond and are therefore sp^2 hybridized.

14.9   (a) 2,3-dimethylhexanal: hexanal means that the base structure is a six-carbon aldehyde. The 2,3-dimethyl denotes that there are two methyl groups connected to this six-carbon chain at C-2 and C-3, thus:

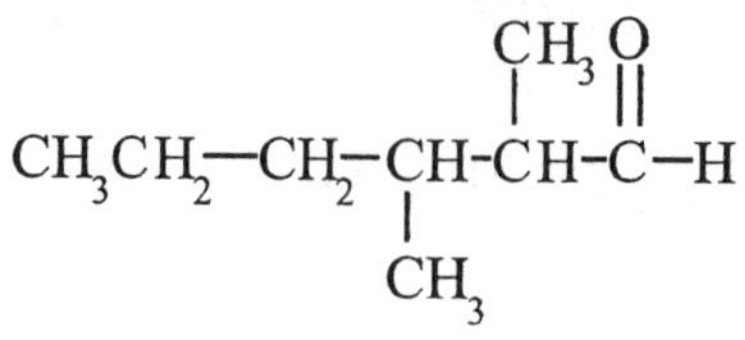

$$CH_3CH_2-CH_2-\overset{\overset{\displaystyle CH_3}{|}}{CH}-\overset{\overset{\displaystyle }{|}}{\underset{\underset{\displaystyle CH_3}{|}}{CH}}-\overset{\overset{\displaystyle O}{||}}{C}-H$$

(b) Ethyl isopropyl ketone: Ketone denotes an organic structure that has two hydrocarbon groups on both sides of a carbonyl C=O. The ethyl denotes that one of the groups is an ethyl group, while the isopropyl denotes that an isopropyl group is on the other side of a carbonyl.

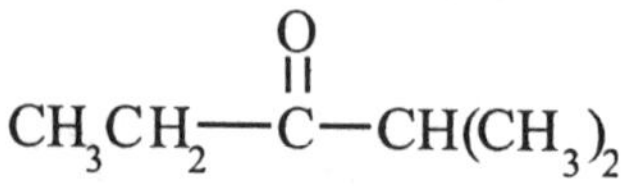

$$CH_3CH_2-\overset{\overset{\displaystyle O}{||}}{C}-CH(CH_3)_2$$

(c) 3-s-butyl-4-ethylbenzaldehyde: benzaldehyde means that the central structure is a benzene ring with an aldehyde group (CHO) attached to it. The carbon that the aldehyde group is attached to is C-1. 4-ethyl denotes that an ethyl group is connected to the benzene ring at C-4, while 3-s-butyl denotes that there is an s-butyl group connected to C-3.

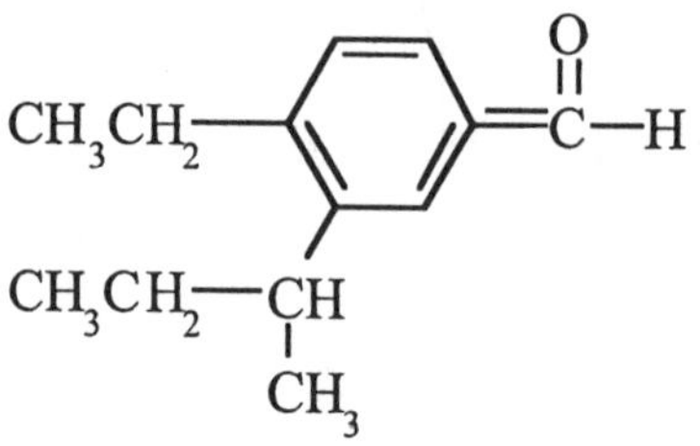

14.11 Method for naming aldehydes:

(i) Find longest chain (or ring) that contains the CH=O. Form the base name by replacing end of alkane name with "al".

(ii) Determine side groups that are attached to this base chain.

(iii) Put names of these groups in alphabetical order at beginning of name. Include numbers to denote which carbon(s) the groups are attached to.

Method for naming Ketones

(0) If the two groups connected to the carbonyl are simple groups, the name is simply the name of the two groups followed by "ketone".

(i) Find longest chain (or ring) that contains the C=O. Form the base name by replacing end of alkane name with "one".

(ii) Count from end to denote which carbon is part of the carbonyl. Be sure to count from end that gives lowest number.

(iii) Determine side groups that are attached to the base chain.

(iv) Put names of these groups in alphabetical order at beginning of name. Include numbers to denote which carbon(s) the groups are attached to.

(a) Using method for naming aldehydes above:

(i) 4 carbons → butanal

(ii) There is an isopropyl group connected to C-2.

(ii) 2-isopropylbutanal

(b) Use method for naming ketones above:

(0) The two groups on the sides of the carbonyl are phenyl and *t*-butyl; thus this compound is named "*t*-butyl phenyl ketone".

(c) Using the method for naming ketones above:

(i) 5-carbon ring → cyclopentanone

(ii) One methyl group attached to C-2.

(iii) 2-methylcyclopentanone

(d) Using method for naming aldehydes above:

(i) 5 carbons → pentanal

(ii) There is a chloro group on C-3 and two methyl groups on C-4.

(iii) 3-chloro-4, 4-dimethylpentanal

14.13   (a) 3-pentanone. Both compounds are ketones and therefore have similar secondary forces, but 3-pentanone has a higher molecular mass.

(b) These two compounds will have similar boiling points because aldehydes and ketones of the same molecular mass have about the same secondary forces.

(c) Aldehydes and ketones of the same molecular mass have about the same water solubility because they both can form hydrogen bonds to H_2O at the carbonyl oxygen.

(d) Butanal. Aldehydes have much stronger secondary attractive forces, dipole-dipole, compared to the weaker dispersion forces of hydrocarbons, and thus a higher boiling point.

(e) Butanal. Aldehydes can hydrogen bond with water, but there are no specific attractive forces between pentane and water.

(f) 2-butanone. Ketones interact via hydrogen bond with water at the carbonyl while the rest of the molecule is essentially a hydrocarbon that does not interact favorably with water. In 2-butanone, the proportion of the molecule that interacts with water is greater than that of 3-pentanone.

(g) 2-butanol. The alcohol, 2-butanol can form strong secondary attractive forces with itself (hydrogen bonds) while the ketone 2-butanone can only form the weaker dipole-dipole interactions.

(h) The water solubilities are very nearly the same because both ketones and alcohols can form hydrogen bonds with water.

14.15   Chromate or permanganate will not oxidize compound 1 because ketones are not oxidized under these mild conditions. Compound 2 will be oxidized to a carboxylic acid (CH_3CH_2COOH) as these mild reagents oxidize aldehydes. Finally, Compound 3 is a primary alcohol that will be oxidized to an aldehyde, which will be further oxidized to a carboxylic acid.

$$CH_3CH_2CH_2CH_2OH \longrightarrow CH_3CH_2CH_2-\overset{\overset{O}{\|}}{C}-H \longrightarrow CH_3CH_2CH_2-\overset{\overset{O}{\|}}{C}-OH$$

14.17   (a) Tollens's reagent will oxidize an aldehyde, but not a ketone or alcohol. A black or shiny mirror precipitate indicates a positive test with Tollens's reagent. Compounds 1 and 2 are not aldehydes and therefore will give a negative result to Tollens's reagent, while compounds 3 and 4 are aldehydes and will give a positive result.

(b) Benedict's reagent will oxidize an α-hydroxy ketone or α-hydroxy aldehyde. Compounds 1 and 4 have an –OH connected to the carbon that is next to the carbonyl group in an aldehyde or ketone and therefore will give a positive test with Benedict's reagent. Compounds 2 and 3 will give negative tests. A red precipitate will indicate a positive test.

14.19   Permanganate will oxidize an aldehyde and a primary alcohol and therefore cannot differentiate between 1-pentanol and pentanal.

14.21   A ketone will undergo reduction in the presence of a metal hydride to form a secondary alcohol, while an aldehyde will also undergo reduction in the presence of a metal hydride to form a primary alcohol, thus:

(a) this compound will undergo reduction to form:

$$\underset{\displaystyle CH_3CH_2 \!\!-\!\! CH\!\cdot\! CH_3}{\overset{\displaystyle OH}{\overset{|}{}}}$$

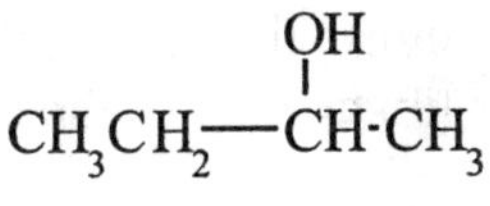

(b) this compound will undergo reduction to form:

$$CH_3CH_2CH_2OH$$

(c) this alcohol will not undergo reduction.

14.23

$$H^- + H_2O \rightarrow H_2 + {}^-OH$$

H^- is a base in this reaction because it accepts a proton.

14.25   In this reaction, the CH_3OH is first added to the C=O double bond to form the hemiacetal, which then undergoes further substitution reaction to form the acetal as:

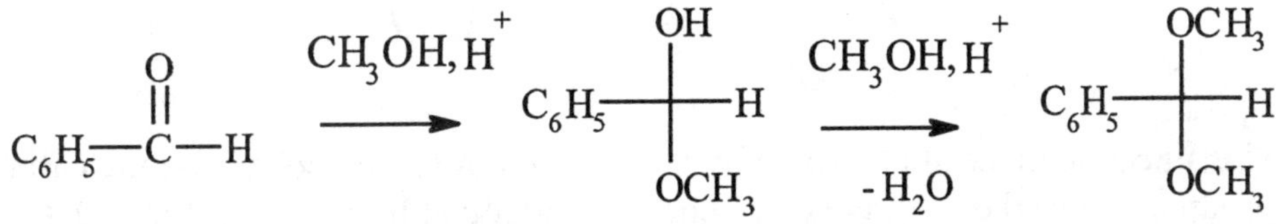

14.27   (a) Formation of a linear acetal occurs by the addition and substitution of an alcohol to a carbonyl group of an aldehyde or ketone. To determine the starting materials of a linear acetal, draw the expanded structure of the acetal, the replace the RO–C–OR linkage with a carbonyl group. The alcohol can then be found by adding an H to the RO– group (ROH). Thus:

$$\underset{\substack{|\\ \text{OCH}_2\text{CH}_3}}{\overset{\substack{\text{OCH}_2\text{CH}_3\\ |}}{\text{C}_6\text{H}_5-\text{C}-\text{H}}}$$

Replace RO–C–OR
with C=O
to get:

$$\text{C}_6\text{H}_5-\overset{\overset{\text{O}}{\|}}{\text{C}}-\text{H}$$

(b) In the formation of a cyclic acetal, a cyclic hemiacetal is first formed from a compound that has an –OH on one end and an aldehyde carbonyl at the other, as shown in Example 14.10. The hemiacetal then reacts with an alcohol to form the acetal. Thus for this structure, the hemiacetal that must first be formed is determined by substituting the ether linkage off of the ring with –OH:

Thus, this hemiacetal must be formed from a four carbon chain with an aldehyde on one end and an –OH on the other.

$$\text{HO}-\text{CH}_2\text{-CH}_2\text{-CH}_2\text{-}\overset{\overset{\text{O}}{\|}}{\text{C}}\text{-OH} \longrightarrow$$

This hemiacetal can then react with methanol to form the acetal as:

CH₃OH

(c) This linear hemiacetal is formed by the addition of an alcohol across the carbonyl group and thus the starting compounds can be found by replacing the HO–C–OR group with a carbonyl to form the starting ketone and recognizing that the ROH must be the starting alcohol:

$$\underset{\substack{|\\ \text{CH}_2\text{CH}_3}}{\overset{\substack{\text{OCH}_2\text{CH}_3\\ |}}{\text{HO}-\text{C}-\text{CH}_2\text{CH}_2\text{CH}_3}}$$

Replace RO–C–OR
with C=O
to get:

$$\text{CH}_2\text{CH}_3-\overset{\overset{\text{O}}{\|}}{\text{C}}-\text{CH}_2\text{CH}_2\text{CH}_3$$

The other starting compound must therefore be the alcohol CH_3CH_2OH.

(d) To complete this exercise, you need to draw out the expanded structure of this acetal, which is:

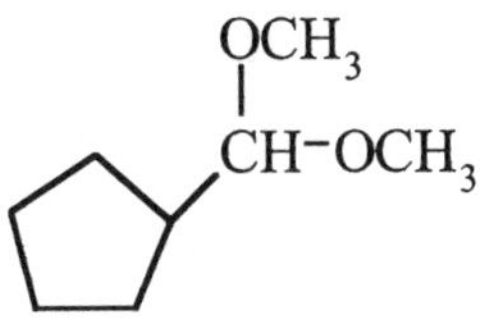

$$
\begin{array}{c}
OCH_3 \\
| \\
CH\text{-}OCH_3
\end{array}
$$

Formation of this acetal occurs by the addition and substitution of an alcohol to a carbonyl group of an aldehyde. To determine the starting materials of this acetal, replace the RO–C–OR linkage with a carbonyl group. The alcohol can then be found by adding an H to the RO– group (ROH). Thus:

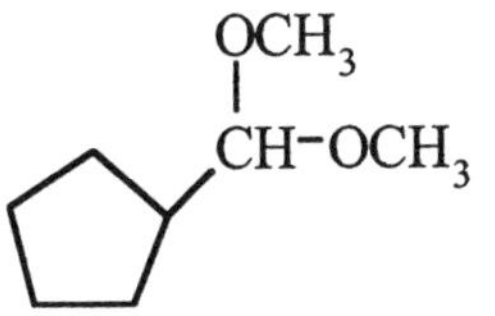

$$
\begin{array}{c}
OCH_3 \\
| \\
CH\text{-}OCH_3
\end{array}
\qquad
\begin{array}{c}
\text{Replace RO–C–OR} \\
\text{with C=O} \\
\text{to get:}
\end{array}
\qquad
\begin{array}{c}
O \\
|| \\
CH
\end{array}
$$

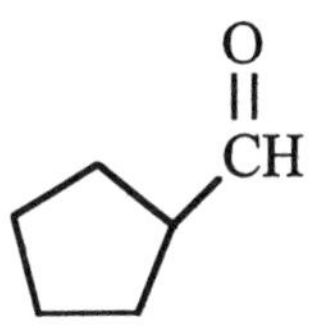

14.29 (a) The carbon and oxygen atoms that are part of C=C and C=O are sp^2 hybridized.

(b) All bond angles in the C=C and C=O double bonds are 120°.

(c) The C=O double bond is polar because it is a bond between two different atoms, while the C=C bond is not.

(d) Both double bonds will undergo a variety of addition reactions, however not exactly the same reaction under similar conditions.

14.31 (a) An aldehyde has the general formula RCHO where R is a hydrocarbon structure. Therefore, in this molecule R = C_4H_9. The one methyl group means that this R group has only one end (i.e., no branching). Therefore, the structure must be:

$$
\begin{array}{c}
O \\
|| \\
CH_3CH_2CH_2CH_2\!-\!CH
\end{array}
$$

(b) This description can only be true if the structure is a five-carbon ring with a C=O connected to it, thus:

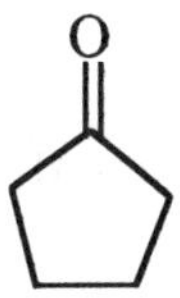

(c) A C_8 ketone denotes that there is a carbonyl group (1 carbon, 1 oxygen) that has two hydrocarbon structures bonded to it. There also exists a benzene ring, which accounts for six of the carbons leaving a single carbon to make up the hydrocarbon structure on the other side of the carbonyl from the benzene ring, thus:

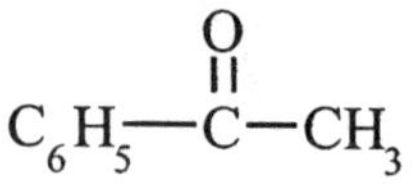

$$C_6H_5-\overset{\overset{\displaystyle O}{\|}}{C}-CH_3$$

(d) $C_6H_{12}O$ → aldehyde = CHO group = 1 carbon and 1 oxygen and 1 hydrogen. A t-butyl group = $C(CH_3)_3$ = 4 carbons and 9 hydrogens. This leaves a single CH_2 group that must also exist in the hydrocarbon group, which must go between the CHO and t-butyl group, thus:

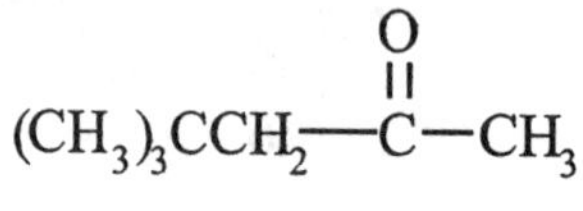

$$(CH_3)_3CCH_2-\overset{\overset{\displaystyle O}{\|}}{C}-CH_3$$

(e) $C_6H_{12}O$ → ketone = a single carbonyl = 1 carbon and 1 oxygen. A t-butyl group = $C(CH_3)_3$ = 4 carbons and 9 hydrogens. This leaves CH_3 for the other hydrocarbon structure on the other side of the carbonyl from the t-butyl group, thus:

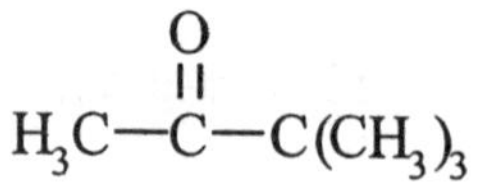

$$H_3C-\overset{\overset{\displaystyle O}{\|}}{C}-C(CH_3)_3$$

14.33 Draw the structure and then rename as in Example 14.11

(a) This structure should be named so that the number that denotes where the methyl group is attached to the ring is the lowest possible. Thus it should be 3-methylcyclopentanone.

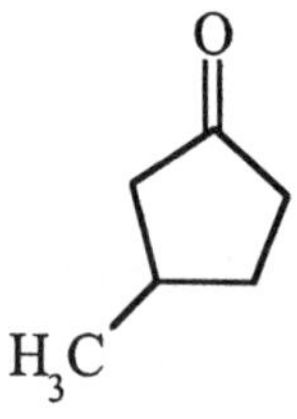

(b) The longest chain in this compound, which contains the carbonyl is five carbons long, not four. Thus the name should be 2-ethylpentanal.

$$\overset{\displaystyle O}{\overset{\displaystyle \|}{CH}}-\underset{\displaystyle \underset{\displaystyle CH_2-CH_2-CH_3}{|}}{CH}-CH_2-CH_3$$

(c) The carbonyl cannot be on C-1 in a ketone, which would make it an aldehyde; thus, the name of this compound is pentanal.

$$\overset{\displaystyle O}{\overset{\displaystyle \|}{CH}}-CH_2-CH_2-CH_2-CH_3$$

(d) The numbering of the carbons where the carbonyl is located and the chloro group is attached should be from the other end of the chain to give 3-chlor-2-pentanone.

$$CH_3-CH_2-\overset{\displaystyle \overset{\displaystyle Cl}{|}}{CH}-\overset{\displaystyle \overset{\displaystyle O}{\|}}{C}-CH_3$$

14.35 (a) 1-butanol. The alcohol, 1-butanol, can form strong secondary attractive forces with itself (hydrogen bonds), while the aldehyde butanal can only form the weaker dipole-dipole interactions. Thus the boiling point of 1-butanol is higher.

(b) Alcohols and ketones of the same molecular mass have about the same water solubility because they both can form hydrogen bonds to H_2O at the carbonyl oxygen.

(c) Butanal. Aldehydes have stronger dipole-dipole attractive forces from the carbonyl than the dipole-dipole from the C–O–C bond of an ether, thus butanal has a higher boiling point.

(d) The water solubilities are very nearly the same because both aldehydes and ethers can form hydrogen bonds with water.

14.37 The compounds that contain an oxygen can form hydrogen bonds to water and will thus be soluble to a similar extent. Alcohols will hydrogen bond slightly better than aldehydes, which will hydrogen bond slightly better than ethers. The hydrocarbon cannot preferentially interact with water and will therefore be least soluble in water.

1-butanol ≥ butanal ≥ diethyl ether > pentane

14.39 Tollens's reagent will oxidize an aldehyde, but not a ketone or alcohol. Therefore, Tollens's reagent will not distinguish between this ketone and tertiary alcohol.

14.41 The aldehyde in glucose will undergo reduction to form the hexol (a compound with six

–OH groups)

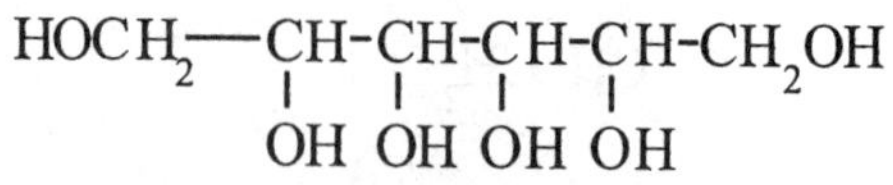

$$HOCH_2-CH-CH-CH-CH-CH_2OH$$
$$\qquad\qquad OH\ OH\ OH\ OH$$

14.43 The reaction of 2-butanone with excess methanol will first form the hemiacetal by addition of the CH_3OH to the carbonyl double bond. The hemiacetal then undergoes further reaction with methanol to form the more stable acetal.

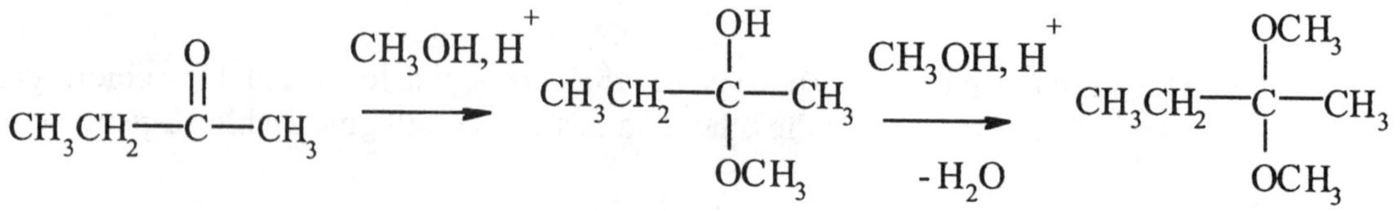

$$CH_3CH_2-\overset{\overset{\displaystyle O}{||}}{C}-CH_3 \xrightarrow{CH_3OH,\,H^+} CH_3CH_2-\overset{\overset{\displaystyle OH}{|}}{\underset{\underset{\displaystyle OCH_3}{|}}{C}}-CH_3 \xrightarrow[-H_2O]{CH_3OH,\,H^+} CH_3CH_2-\overset{\overset{\displaystyle OCH_3}{|}}{\underset{\underset{\displaystyle OCH_3}{|}}{C}}-CH_3$$

14.45 (a) This ketone will react with excess methanol to form the acetal:

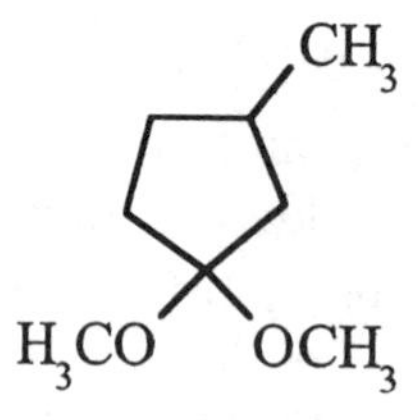

(b) This ketone will undergo reduction in the presence of H_2 and Ni to form:

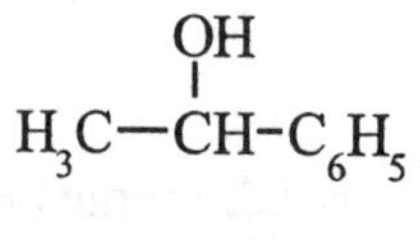

$$H_3C-\overset{\overset{\displaystyle OH}{|}}{CH}-C_6H_5$$

(c) An acetal cannot undergo a reduction reaction, thus there is no reaction.

(d) Tollens's reagent will oxidize an aldehyde to a carboxylic acid to form:

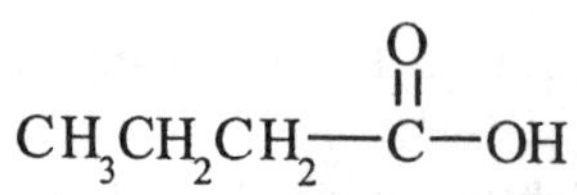

$$CH_3CH_2CH_2-\overset{\overset{\displaystyle O}{||}}{C}-OH$$

(e) An acetal will not undergo an oxidation reaction in the presence of Benedict's reagent, thus there is no reaction.

(f) NADH acts similar to a metal hydride to reduce the aldehyde to a primary alcohol:

$$HOCH_2CH_2C_2H_5$$

(g) A ketone will undergo reduction in the presence of a metal hydride and subsequent addition of water to form:

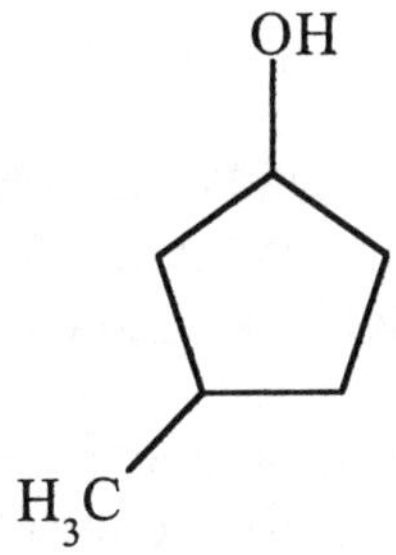

(h)

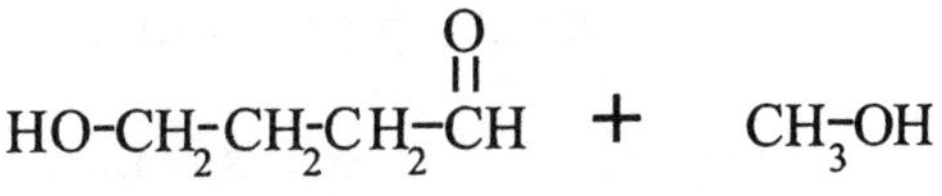

$$HO\text{-}CH_2\text{-}CH_2\text{-}CH_2\text{-}CH \quad + \quad CH_3\text{-}OH$$

14.47 Reduction is equivalent to adding two hydrogens across a double bond, thus:

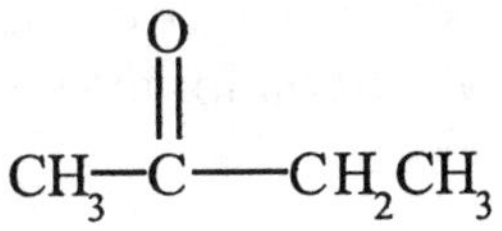

$$CH_3\text{-}C\text{---}CH_2CH_3$$

14.49 (a) A tertiary alcohol cannot be oxidized to form an aldehyde or ketone.

(b) This primary alcohol will be oxidized to form the aldehyde:

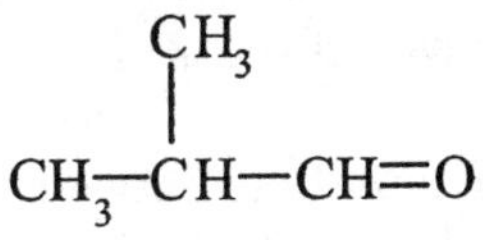

$$CH_3\text{-}CH\text{-}CH{=}O$$ with CH₃ branch

$$\begin{array}{c} CH_3 \\ | \\ CH_3\text{-}CH\text{-}CH{=}O \end{array}$$

(c) This secondary alcohol will be oxidized to form the ketone:

(d) Phenol will not undergo oxidation.

14.51 Vanillin is a phenol and therefore will be acidic enough to turn blue litmus paper red, while the ketone menthone will be neutral to litmus. Vanillin is also an aldehyde that will show a positive Tollens's test while the ketone will show a negative Tollens's test.

14.53 The hydride reduction only reacts with the carbonyl C=O; however, the catalytic hydrogenation will react with the C=O bonds and the C=C double bond in cinnamaldehyde, thus losing two more hydrogens.

14.55 The cyclic compounds are formed by the reaction of the –OH at C-5 with the carbonyl at C-1 to form a stable cyclic hemiacetal. There are two structures because there are two positions (up and down) for the –OH group on C-1 of the hemiacetal.

14.57 Counting up the carbons, oxygens, and hydrogens in the structure in Box 14.1 gives $C_{21}H_{30}O_2$.

14.59 There are two pentanone isomers, 2 pentanone and 3-pentanone. The number in the name is needed to distinguish between the two isomers. There is only one butanone compound because the carbonyl carbon for the unbranched C_4 ketone must be the next to last carbon in the four-carbon chain.

14.61 The number of peaks in the ^{13}C NMR corresponds to the number of unique carbons in the structure. Six peaks in the ^{13}C NMR means the unknown has six unique carbons. Out of the three structures, only B has 6 unique carbons; the aldehyde carbon, the methyl group carbon, the carbon of the benzene ring that is bonded to the methoxy, the aromatic carbon bonded to that carbon (there are two of these), the carbon that is bonded to the aldehyde, and the aromatic carbon bonded to this one (there are two of these). Both A and C have eight unique carbons, and thus would show eight peaks in a ^{13}C NMR.

14.63 Yes, IR spectroscopy would differentiate between progesterone and testosterone. Testosterone would show a peak at 3200–3650 cm^{-1} corresponding to the alcohol O–H.

14.65 The reaction of V with H_2/Pt to yield W, which adds two hydrogens across a double bond. A compound with the formula C_4H_8O that contains a double bond is butanal, which adds hydrogens across the C=O bond to form 1-butanol, $C_4H_{10}O$. The alcohol 1-butanol can be heated in the presence of an acid catalyst to remove water, H_2O and form 1-butene (compound X, C_4H_8). The unsaturated hydrocarbon 1-butene can then react with water in an acidic environment to form 2-butanol (compound Y). This alcohol will oxidize to butanone, compound Z).

14.67

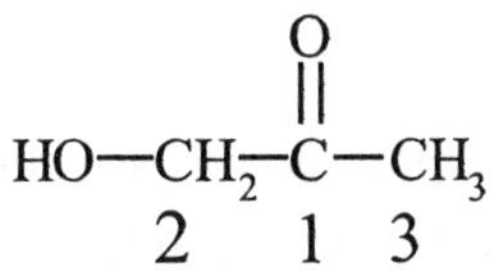

$$\text{HO}\!-\!\overset{}{\underset{2}{\text{CH}_2}}\!-\!\overset{\overset{\displaystyle O}{\|}}{\underset{1}{\text{C}}}\!-\!\underset{3}{\text{CH}_3}$$

14.69   Biological processes occur in an aqueous environment, and LiAlH$_4$ is not compatible with an aqueous environment, while NADH is compatible. LiAlH$_4$ will preferentially react with water instead of the carbonyl group, while NADH does not react with water.

Chapter 15

Carboxylic Acids, Esters, and Other Acid Derivatives

15.1 A carboxylic acid is a compound with a carbonyl bonded to a hydroxyl group; an ester is a compound with a carbonyl bonded to an oxygen atom; an amide group is a compound that has a carbonyl group that is bonded to an amino group (a nitrogen atom); an anhydride is a compound that has a carbonyl group bonded to two oxygen atoms; and, finally, an acid halide is a compound that has a carbonyl group bonded to a halogen (Cl, F, I...), thus:

Compound 1 is none of the possibilities listed, it is a ketone and an ether because the two oxygen atoms are not bonded to the same carbon. Compound 2 is an ester; compound 3 is an acid chloride; and compound 4 is none of the possibilities listed, it is a ketone and an alcohol because the two oxygen atoms are not bonded to the same carbon. Compound 5 has a carbonyl bonded to a hydroxyl group and is thus a carboxylic acid; compound 6 is an amide as it has a carboxyl group bonded to a nitrogen; compound 7 is carboxylic acid as it has a carbonyl bonded to a hydroxyl group, this compound is also an ether as it has an oxygen bonded to two sp^3 carbons; Compound 8 is an anhydride because it has a carbonyl group bonded to two oxygen atoms.

15.3 A carboxylic acid has the structure R-COOH, where R denotes a hydrocarbon structure. This molecule has the formula $C_5H_{10}O_2$, thus the R group must consist of 4 carbons and 9 hydrogens. The isomers that can be formed from these combinations are:

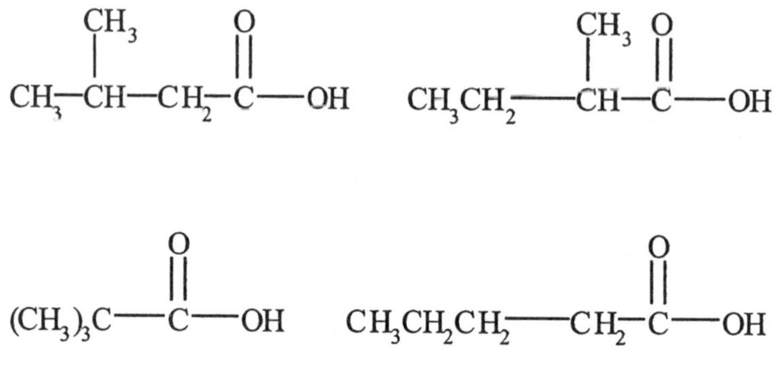

15.5 Carboxylic acids are synthesized by the selective oxidation of primary alcohols, aldehydes, or an alkyl benzene, thus:

(a) In this alkyl benzene, isopropyl benzene, the alkyl group will be oxidized to a carboxylic acid:

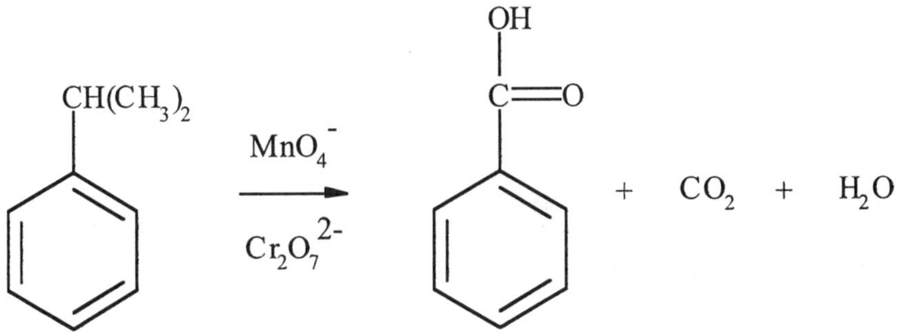

(b) This primary alcohol will first oxidize to an aldehyde and then to a carboxylic acid:

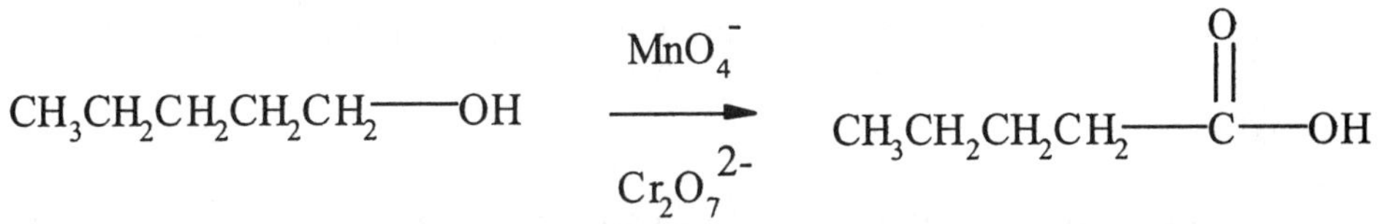

(c) 2-propanol is a secondary alcohol that will oxidize to a ketone, which will not undergo further oxidation.

15.7   Method for naming carboxylic acids:

   (i)   Find longest chain (or ring) that contains the COOH. Form the base name by replacing end of alkane name with "oic acid".

   (ii)   Determine side groups that are attached to this base chain.

   (iii) Put names of these groups in alphabetical order at beginning of name. Include numbers to denote which carbon(s) the groups are attached to.

   (a)

      (i) Longest chain = 6 carbons = hexanoic acid.

      (ii) There is an isopropyl group on C-2 and two methyl groups on C-2 and C-5.

      (iii) 2-isopropyl-2, 5-dimethylhexanoic acid.

   (b)

      (i) Longest chain = 5 carbons = pentanoic acid.

      (ii) There is a chloro group on C-3 and two methyl groups on C-4.

(iii) 3-chloro-4, 4-dimethylpentanoic acid.

(c)

(i) Longest chain = 4 carbons = butanoic acid.

(ii) There is a phenyl group on C-3.

(iii) 3-phenylbutanoic acid.

(d)

(i) Longest chain that contains both carboxyl groups is 3 carbons = propanedioic acid.

(ii) There is no need to number the carbons of the carboxyl because these must be at the ends of the longest chain. The group between the two carboxyls is an ethyl group. There is no need to number this, as it must be on the 2 carbon.

(iii) ethylpropanedioic acid.

15.9   A pair of carboxylic acid molecules can align to form a dimer that allows two hydrogen bonds between the two molecules as:

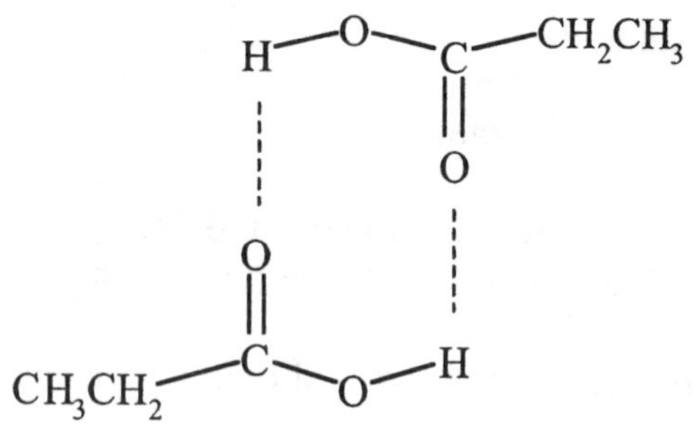

15.11   (a) Two carboxylic acid molecules can form hydrogen bonds that can align to form a dimer that allows two hydrogen bonds between the two molecules. This "double" hydrogen bonding is a stronger intermolecular interaction than the single hydrogen bond that can form in alcohols, thus ethanoic acid has a higher boiling point than 1-propanol.

(b) The solubilities of these two compounds are about the same because they can both form similar hydrogen bonds with water.

15.13   When the –OH is directly bonded to the carbonyl, the molecule is a carboxylic acid and more acidic than the alcohol and ketone of compound 2.

15.15   If a compound that is added to a carboxylic acid is a base, an acid–base reaction will occur.

(a) Water is a weak base and thus a reaction will occur, but the equilibrium will be far to the left.

$$CH_3-\overset{\overset{\textstyle O}{\|}}{C}-OH \quad + \quad H_2O \quad \longrightarrow \quad CH_3-\overset{\overset{\textstyle O}{\|}}{C}-O^- \quad + \quad H_3O^+$$

(b) NaOH is a strong base, thus:

$$CH_3-\overset{\overset{\textstyle O}{\|}}{C}-OH \quad + \quad NaOH \quad \longrightarrow \quad CH_3-\overset{\overset{\textstyle O}{\|}}{C}-O^-Na^+ \quad + \quad H_2O$$

(c) NaHCO$_3$ is a base, thus:

$$CH_3-\overset{\overset{\textstyle O}{\|}}{C}-OH \quad + \quad NaHCO_3 \quad \longrightarrow \quad CH_3-\overset{\overset{\textstyle O}{\|}}{C}-O^-Na^+ \quad + \quad CO_2 \quad + \quad H_2O$$

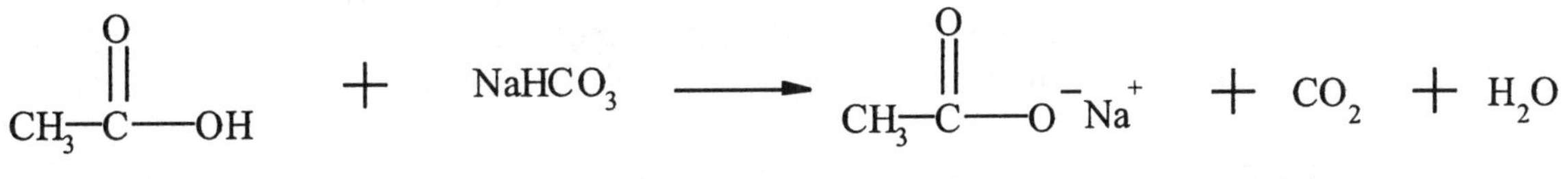

(d) Water is a weak base and thus a reaction with phenol will go, but the equilibrium will be far to the left.

$$C_6H_5-OH \quad + \quad H_2O \quad \longrightarrow \quad C_6H_5-O^- \quad + \quad H_3O^+$$

(e) NaOH is a strong base that will react with phenol, which is slightly acidic, thus:

$$C_6H_5-OH \quad + \quad NaOH \quad \longrightarrow \quad C_6H_5-O^- \quad + \quad Na^+ \quad + \quad H_2O$$

15.17 Propanoic acid dominates in water because the extent of ionization is less than 2%. To form a solution that has propanoic acid and a pH, a substantial amount of buffer must also be added to the solution. The presence of this buffer causes the propanoate ion to be the dominant species.

15.19   Propanoic acid is acidic and will turn blue litmus paper red, while the alcohol propanol will not. Thus the unknown is propanoic acid.

15.21   (a) Sodium hexanoate: hexanoate denotes that the carboxylic acid from which the salt is derived has six carbons. Sodium denotes that the hydrogen of the –COOH is replaced by sodium. The sodium ion has a charge of +1, thus:

$$CH_3CH_2CH_2CH_2CH_2 \overset{\overset{\displaystyle O}{\|}}{-C} -ONa$$

(b) Calcium benzoate: benzoate denotes that the carboxylic acid from which the salt is derived is benzoic acid. Calcium denotes that the hydrogen of the –COOH is replaced by calcium. Calcium ion has a +2 charge, thus:

$$(C_6H_5COO)_2Ca$$

15.23   (a) Carboxylate salts are ionic compounds while carboxylic acids are covalent. The ionic intermolecular interactions of an ionic compound are much stronger than any intermolecular interactions of covalent compounds, even hydrogen bonding. Thus, sodium propanoate has a higher melting point than propanoic acid.

(b) Carboxylate salts are ionic compounds that dissociate completely in water. The interaction between these dissociated ions and water is much stronger than the hydrogen bond interaction between a carboxylic acid and water; thus sodium hexanoate will be more soluble in water than octanoic acid.

(c) The polar covalent octanoic acid will interact more strongly with hexane, a nonpolar alkane, than the ionic carboxylate salt, sodium hexanoate.

15.25   If a compound turns red litmus paper blue, that signifies that the compound is a base. Butanoic acid is an acid while sodium butanoate is a base; thus the compound must be sodium butanoate.

15.27   A soap is a compound that has a molecular structure such that part of the molecule will prefer an organic environment while another portion of the molecule will preferentially interact with water. This usually occurs by having one end of the molecule be an ionic compound while the other is a long hydrocarbon tail. Of the structures that are listed in this problem, only compound 3 fits this description.

15.29   An ester will be formed by the reaction of an alcohol and a carboxylic acid in an acidic environment. The overall result is that the acid loses an –OH from the –COOH group while the alcohol loses an –H from the –OH group. The resulting carbonyl and –O groups then link to form the ester. Thus these compound react as:

(a)

$$CH_3CH_2CH_2CH_2 \overset{\overset{\displaystyle O}{\|}}{-C} -OH \quad + \quad HOCH_2CH_3 \quad \longrightarrow \quad CH_3 \overset{\overset{\displaystyle O}{\|}}{-C} -O-CH_3$$

(b)

$$C_6H_5 \overset{\overset{\displaystyle O}{\|}}{-C} -OH \quad + \quad HOCH_2CH_2CH_2CH_3 \quad \longrightarrow \quad CH_3 \overset{\overset{\displaystyle O}{\|}}{-C} -O-CH_3$$

15.31 To determine the starting materials from which an ester is formed, it is best to break the ester linkage between the carbonyl and the oxygen to form two compounds. Then make the starting carboxylic acid from the compound with the carbonyl and add an –OH group to it, while the starting alcohol is formed by adding an H atom to the compound that has the –O endgroup, thus:

(a) 4,4-dimethylpentanoic acid and 2-propanol

(b) *p*-methylbenzoic acid and 2-butanol

(c) butanoic acid and *p*-methylphenol

15.33 (a) *t*-butyl butanoate: butanoate denotes that the carboxylic acid from which the ester is formed is butanoic acid. In other words the hydrocarbon group that is directly bonded to the carbonyl is a four carbon straight chain (butane). *t*-Butyl denotes that the alcohol from which the ester is derived is *t*-butyl alcohol. In other words, the hydrocarbon group that is directly bonded to the –O of the ester group is a *t*-butyl group, thus:

$$CH_3CH_2CH_2 \overset{\overset{\displaystyle O}{\|}}{-C} -O-C(CH_3)_3$$

(b) Cyclohexyl benzoate: benzoate denotes that the carboxylic acid from which the ester is formed is benzoic acid. In other words the hydrocarbon group that is directly bonded to the carbonyl is a benzene ring. Cyclohexyl denotes that the alcohol from which the ester is derived is cyclohexanol. In other words, the hydrocarbon group that is directly bonded to the –O of the ester group is a cyclohexene ring, thus:

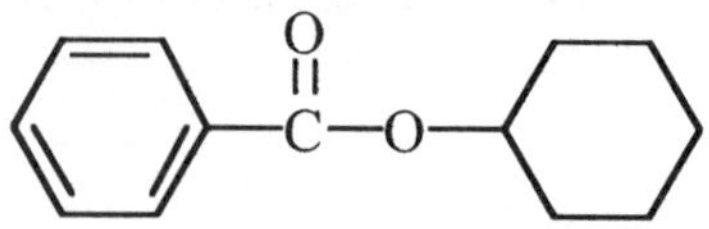

15.35 (a) Both methyl propanoate and pentanal have carbonyl groups that can form dipole-dipole interactions of similar strength. However the carbonyl group in the ester, methyl propanoate, is in the middle of the molecule while the carbonyl group of the aldehyde pentanal is at the end of the molecule. Therefore it will be easier for two dipoles of the aldehyde to come together to form an intermolecular interaction than for the same thing to happen between two ester molecules. Therefore, pentanal will have more secondary interactions and thus a higher boiling point.

(b) Water can just as easily hydrogen bond with the carbonyl in an ester as with the carbonyl of an aldehyde; therefore these two molecules will have similar solubilities in water.

(c) The hydrogen bonding in 1-pentanol is a stronger secondary force than the dipole-dipole interactions of the ester methyl propanoate. Thus 1-pentanol will have the higher boiling point.

(d) Water can hydrogen bond with the carbonyl or –OH group in a carboxylic acid, while it can only hydrogen bond to the carbonyl group of an ester. Therefore, hexanoic acid will be more soluble in water than methyl pentanoate.

(e) The hydrogen bonding in butanoic acid is a stronger secondary force than the dipole-dipole interactions of the ester methyl propanoate. Thus butanoic acid will have the higher boiling point.

15.37 Polyesters can be formed by the reaction of diacids with diols, similar to the way esters are formed from acids and alcohols. Therefore, the diol will be a compound HO–R–OH and the diacid will have the structure HOOC–R'–COOH. The exact structure of R and R' from a polyester structure can be found by noticing that R is the hydrocarbon group that is between the two oxygen atoms that are part of the carbonyl, while R' is the hydrocarbon structure that is between the carbonyls. Thus, in this compound R = $CH_2CH_2CH_2$ and R' = $CH_2CH_2CH_2CH_2$. Therefore, this polyester is synthesized from 1,6 hexanedioic acid and 1,3-propanediol.

15.39 The ester linkage can be broken under acidic conditions to form the original acid and alcohol or under basic conditions to form a carboxylate salt and an alcohol (or phenol). Thus, these compounds undergo the following reactions:

(a)

$$CH_3CH_2-\overset{\overset{\displaystyle O}{\|}}{C}-OCH_2CH_2CH_3 \quad \xrightarrow[\text{H}_2\text{O}]{\text{H}^+} \quad CH_3CH_2-\overset{\overset{\displaystyle O}{\|}}{C}-OH \quad + \quad HO-CH_2CH_2CH_3$$

(b)

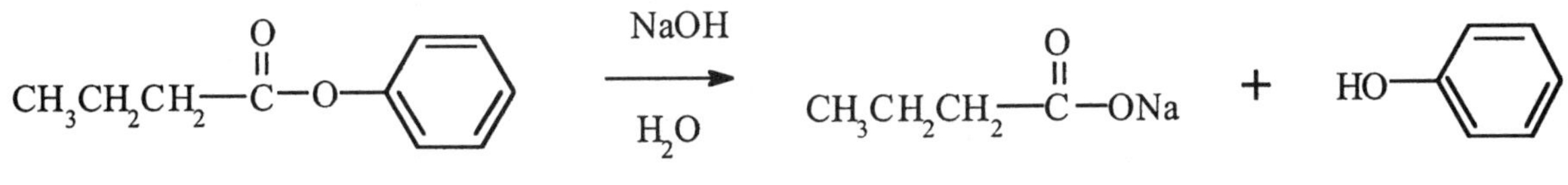

$$CH_3CH_2CH_2-\overset{\overset{\displaystyle O}{\|}}{C}-O-\bigcirc \quad \xrightarrow[\text{H}_2\text{O}]{\text{NaOH}} \quad CH_3CH_2CH_2-\overset{\overset{\displaystyle O}{\|}}{C}-ONa \quad + \quad HO-\bigcirc$$

15.41 (a) Acid halides are compounds where the –OH of a carboxylic acid is replaced by a halogen and are named by replacing the "ic acid" with "yl halide". Thus for this compound, the –OH of 2-methylpropanoic acid is replaced by a chlorine to form:

$$(CH_3)_2CH-\overset{\overset{\displaystyle O}{\|}}{C}-Cl$$

(b) An anhydride is a compound that has two carbonyl groups that are bonded to a single oxygen molecule as R–(C=O)–O–(C=O)–R. The R structure is denoted by the name of the anhydride, i.e., ethanoic denotes that the R groups are CH_3 in this structure, thus:

$$H_3C-\overset{\overset{\displaystyle O}{\|}}{C}-O-\overset{\overset{\displaystyle O}{\|}}{C}-CH_3$$

15.43 Acid chlorides react with water to form the corresponding carboxylic acid as:

$$C_6H_5-\overset{\overset{\displaystyle O}{\|}}{C}-Cl \quad \xrightarrow[\text{H}_2\text{O}]{-\text{HCl}} \quad C_6H_5-\overset{\overset{\displaystyle O}{\|}}{C}-OH$$

15.45 The HO–P=O bond in phosphoric acid behaves similarly to a carboxylic acid group by undergoing esterification reaction with alcohols. Thus a phosphoric ester will be formed by the reaction of an alcohol and a phosphoric acid. The overall result is that the acid loses an –OH from the –POOH group while the alcohol loses an –H from the –OH group.

The resulting –P=O and –O groups then link to form the ester. Thus these compounds react as:

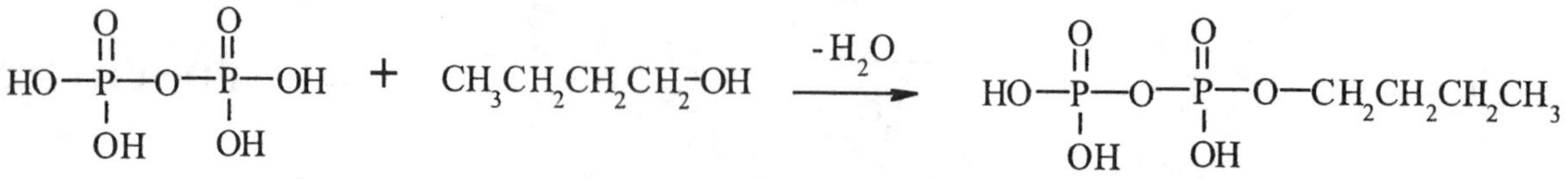

This compound is butyl diphosphate.

15.47   Diphosphoric acid will give up its hydrogens that are part of an –OH group in the presence of a base. The only way that a solution of an acid can have a pH of 7 is if there is also a base present to buffer the action of the acid, therefore:

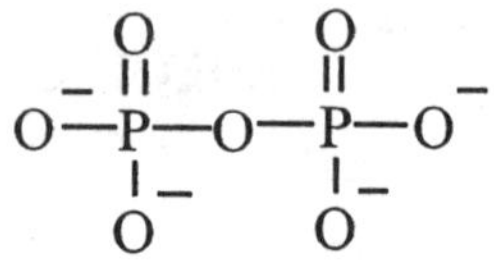

15.49   *cis*-2-methylcyclopentane carboxylic acid: 2-methyl cyclopentane denotes a five-membered ring with a methyl group attached to the ring at C-2. The carboxylic acid denotes that there is also a –COOH group attached to the ring at C-1, thus:

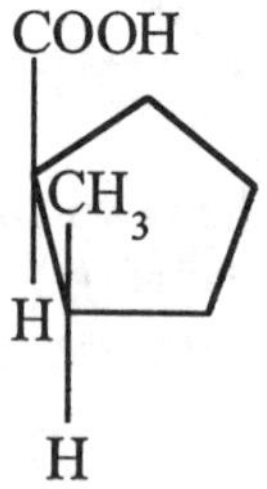

15.51   5-hydroxypentanoic acid is:

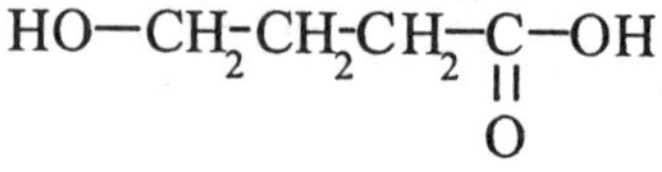

$$HO—CH_2CH_2CH_2—\underset{\underset{O}{\|}}{C}—OH$$

Thus, if the –OH group on one end wraps around to react with the acid group on the other end to form an ester linkage it will form a six-membered ring (including the oxygen atom) as:

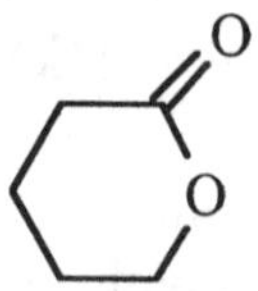

15.53 (a) *trans*-3-chlorocyclohexanecarboxylic acid: cyclohexane denotes that the base structure is a six-membered ring. Carboxylic acid at the end denotes that there is a –COOH group attached to the ring at C-1. *trans*-3-chloro denotes that there is a chloro group on C-3 on the opposite side of the ring from the COOH group:

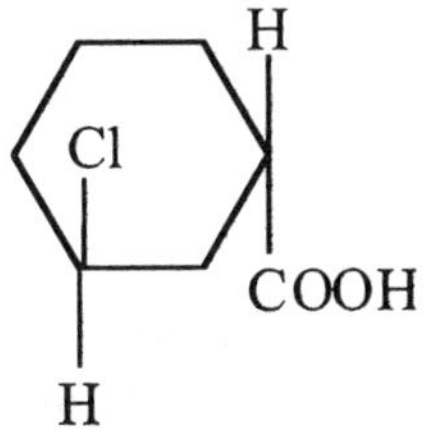

(b) Calcium acetate: acetate denotes the carboxylate salt of acetic acid, CH_3COOH. Calcium is the counter ion. Because calcium has a +2 charge and the acetate ion has a –1 charge, there must be two acetates for each calcium ion.

$$(CH_3COO)_2Ca$$

(c) Isopropyl benzoate: This compound is an ester with R–COO–R'. R' is the hydrocarbon structure denoted by the first group in the name, isopropyl, while R is the base structure of the carboxylic acid from which the ester can be formed. This is denoted in the name as the base name of the second group in the name benzoate → benzoic acid, thus:

$$C_6H_5\text{—}\overset{\displaystyle O}{\overset{\|}{C}}\text{—}O\text{—}CH(CH_3)_2$$

(d) Pentanoic anhydride: An anhydride is a compound that has two carbonyl groups that are bonded to a single oxygen molecule as R–(C=O)–O–(C=O)–R. The R structure is denoted by the name of the anhydride, i.e., pentanoic denotes that the R groups are $CH_3CH_2CH_2CH_2$ in this structure, thus:

$$CH_3CH_2CH_2CH_2\text{—}\overset{\displaystyle O}{\overset{\|}{C}}\text{—}O\text{—}\overset{\displaystyle O}{\overset{\|}{C}}\text{—}CH_2CH_2CH_2CH_3$$

(e) Propanoyl chloride: "yl chloride" denotes an acid chloride where the –OH of a carboxylic acid is replaced by a –Cl. Propanoyl denotes that the base hydrocarbon has three carbons, thus:

$$CH_3CH_2\text{—}\overset{\displaystyle O}{\overset{\|}{C}}\text{—}Cl$$

(f) Methyldiphosphate: Diphosphate denotes a phosphoric acid ester with two phosphorous atoms. Methyl denotes that one of the hydrogens of an –OH group is replaced by a methyl group, thus:

$$\begin{array}{ccc} & O & & O \\ & \parallel & & \parallel \\ HO-&P&-O-&P-OCH_3 \\ & \mid & & \mid \\ & OH & & OH \end{array}$$

(g) Dimethyl phosphate: phosphate denotes a phosphoric acid ester with one phosphorous atom. The dimethyl denotes that two of the hydrogens are replaced with CH_3 groups as:

$$\begin{array}{c} O \\ \parallel \\ HO-P-OCH_3 \\ \mid \\ OCH_3 \end{array}$$

15.55 Atoms a and c are sp^3 hybridized while atoms b and d are sp^2 hybridized. The angles A and C are about the carbon of a carbonyl and are thus 120°, while angle B is about a C–O–C bond which is close to 109.5°.

15.57 If a compound turns blue litmus paper red, that signifies that the compound is an acid. Of these three molecules, both phenol and propanoic acid are strong enough acids to turn blue litmus paper red, while propanol is not. Thus the unknown is either phenol or propanoic acid.

15.59 An ester will be formed by the reaction of a thiol and a carboxylic acid in an acidic environment. The overall result is that the acid loses an –OH from the –COOH group while the alcohol loses an –H from the –SH group. The resulting carbonyl and –S groups then link to form the ester. Thus, these compounds react as:

$$\begin{array}{c} O \\ \parallel \\ CH_3\!-\!C\!-\!OH \end{array} \quad + \quad HS\!-\!CH_2CH_2CH_2CH_3 \quad \xrightarrow[-H_2O]{H^+} \quad \begin{array}{c} O \\ \parallel \\ CH_3\!-\!C\!-\!SCH_2CH_2CH_2CH_3 \end{array}$$

15.61 (a) An ester will not react with water if there is no acid present to catalyze the reaction, thus there is no reaction under these conditions.

(b) An ester will hydrolyze in the presence of water and acid to form an alcohol and carboxylic acid:

$$\begin{array}{c} O \\ \parallel \\ CH_3CH_2\!-\!C\!-\!OH \end{array} \quad + \quad HOCH_2C_6H_5$$

(c) An ester will form an alcohol and carboxylate salt in the presence of water and a base as:

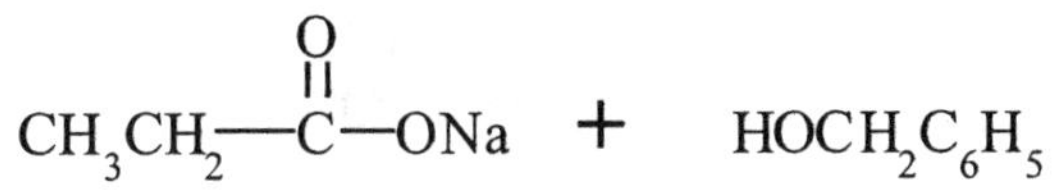

$$CH_3CH_2—\overset{\overset{\displaystyle O}{\|}}{C}—ONa \quad + \quad HOCH_2C_6H_5$$

(d) A lactone is an ester that will also undergo a reaction to form an alcohol and a carboxylate salt. However because the initial compound is a ring, they will be on opposite ends of the same molecule as:

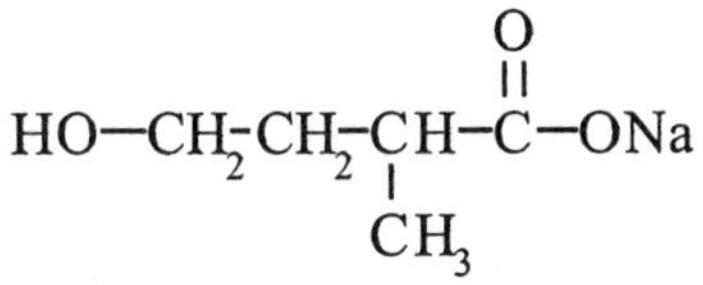

$$HO—CH_2CH_2\underset{\underset{\displaystyle CH_3}{|}}{CH}-\overset{\overset{\displaystyle O}{\|}}{C}—ONa$$

(e) A phosphoric acid will react with an alcohol to form the phosphoric ester as:

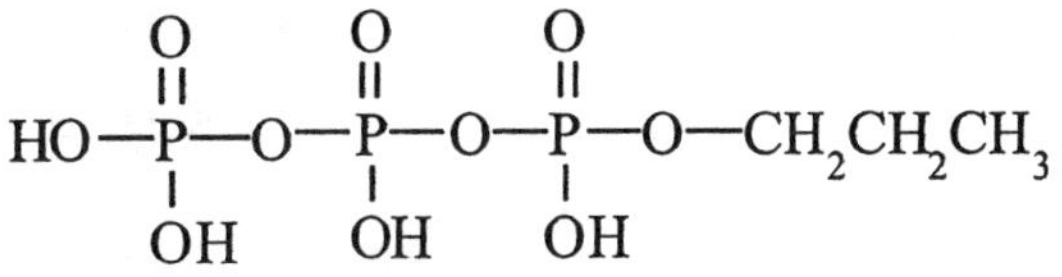

$$HO—\underset{\underset{\displaystyle OH}{|}}{\overset{\overset{\displaystyle O}{\|}}{P}}—O—\underset{\underset{\displaystyle OH}{|}}{\overset{\overset{\displaystyle O}{\|}}{P}}—O—\underset{\underset{\displaystyle OH}{|}}{\overset{\overset{\displaystyle O}{\|}}{P}}—O—CH_2CH_2CH_3$$

(f) A carboxylic acid will undergo an acid base reaction in the presence of a base and will form a carboxylate salt as:

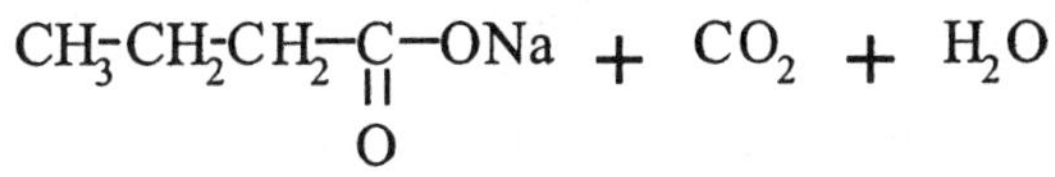

$$CH_3CH_2CH_2\underset{\underset{\displaystyle O}{\|}}{C}—ONa \quad + \quad CO_2 \quad + \quad H_2O$$

(g) An alcohol and a carboxylic acid will react under acidic conditions to form an ester as:

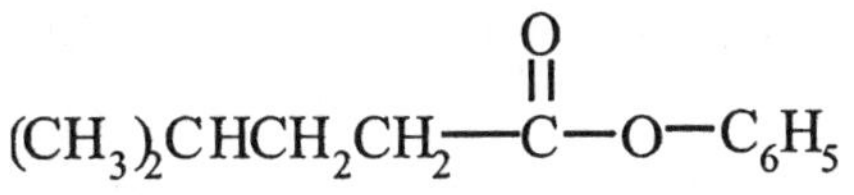

$$(CH_3)_2CHCH_2CH_2—\overset{\overset{\displaystyle O}{\|}}{C}—O—C_6H_5$$

(h) An anhydride will react with water to form a carboxylic acid:

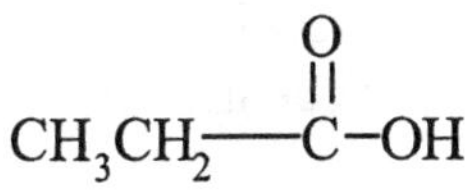

$$CH_3CH_2—\overset{\overset{\displaystyle O}{\|}}{C}—OH$$

(i) An acid chloride will react with water to form a carboxylic acid:

$$CH_3CH_2 \overset{\overset{\displaystyle O}{\displaystyle \|}}{-}C-OH$$

(j) A thio-ester will undergo a reaction in the presence of water and acidic conditions to form a thiol and a carboxylic acid:

$$CH_3CH_2 \overset{\overset{\displaystyle O}{\displaystyle \|}}{-}C-OH \quad + \quad HS-CH_2C_6H_5$$

15.63 A carboxylic acid is acidic, while an ester is not. Therefore if a drop of the unknown is placed on blue litmus paper, if it is an ester it will not turn the paper red. However if the litmus paper turns red, then the unknown is a carboxylic acid.

15.65

$$
\begin{array}{c}
CH_2-OH \\
| \\
CH-OH \\
| \\
CH_2-OH
\end{array}
\quad + \quad 3\ HO-NO_2 \quad \longrightarrow \quad
\begin{array}{c}
CH_2-O-NO_2 \\
| \\
CH-O-NO_2 \\
| \\
CH_2-O-NO_2
\end{array}
$$

15.67 As in Box 9.1, if $[H_3O]^+ = x$, $x^2 = 1.74 \times 10^{-5}$; $x = 4.17 \times 10^{-3}$. $pH = -\log[H_3O]^+ = 2.38$

15.69 Electrical conductivity is proportional to the amount of ions in solution. Because sodium acetate is a salt, it separates 100% into ions in water, while acetic acid is only weakly ionized. This means that there are less ions in the acetic acid solution, and thus its conductivity is significantly lower than that of a solution of sodium acetate.

15.71 3-methylbutyl ethanoate indicates that this ester is a 3-methylbutyl group bonded to the oxygen and two carbons (the carbonyl carbon and a methyl group) on the other side of the oxygen. Thus, the reaction to make this compound is:

$$CH_3 \overset{\overset{\displaystyle O}{\displaystyle \|}}{-}C-OH + HO \cdot CH_2CH_2CH(CH_3)_2 \quad \xrightarrow[-H_2O]{H^+} \quad CH_3 \overset{\overset{\displaystyle O}{\displaystyle \|}}{-}C-OCH_2CH_2CH(CH_3)_2$$

15.73 The NaOH converts all of the acetic acid to sodium acetate, which is 100% ionized, so C will have the greatest concentration of acetate ion. The HCl suppresses the ionization of the acetic acid as it competes to form hydronium ions, thus this will have the least amount of acetate ion. Thus the correct answer is $C > A > B$.

15.75 Hydroxybutanedioic acid has the molecular formula $C_3H_6O_2$. As the isomer is a carboxylic acid, one carbon, two oxygens, and one hydrogen will be used in the formation of the –COOH group, leaving C_2H_5, an ethyl group connected to this carboxylic acid group. Thus the structure is:

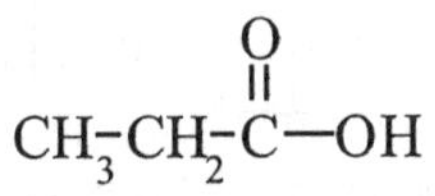

$$\underset{CH_3^-}{} CH_2^- \overset{\overset{\displaystyle O}{\|}}{C}-OH$$

15.77 The alcohol must be converted to carboxylic acids by oxidation. Two carboxylic acids can then react to form the ester group in acidic conditions:

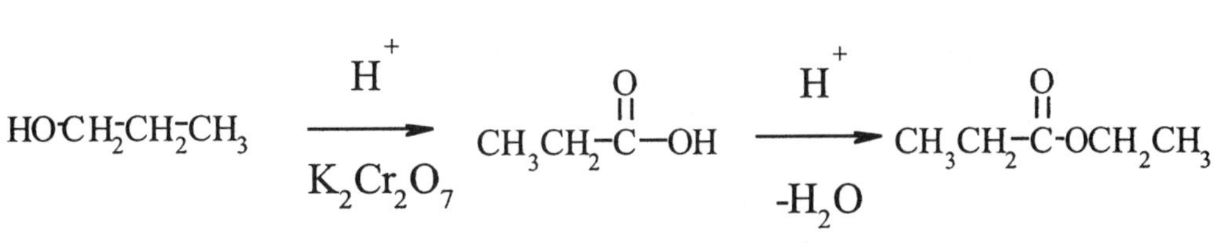

$$HO\text{-}CH_2CH_2CH_3 \quad \xrightarrow[K_2Cr_2O_7]{H^+} \quad CH_3CH_2\overset{\overset{\displaystyle O}{\|}}{C}-OH \quad \xrightarrow[-H_2O]{H^+} \quad CH_3CH_2\overset{\overset{\displaystyle O}{\|}}{C}\text{-}OCH_2CH_3$$

15.79 The carboxylic acid group in *p*-aminobenzoic acid and the alcohol of ethanol will react to form an ester to form benzocaine:

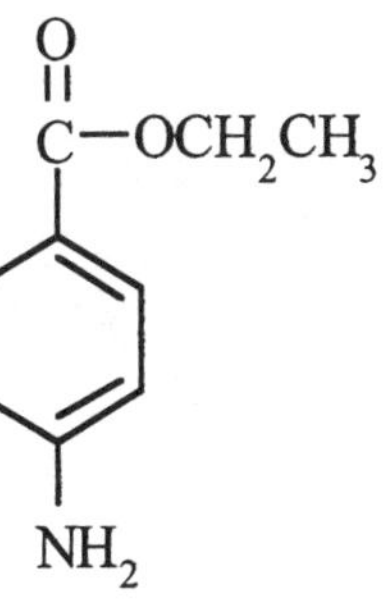

15.81 The presence of the chlorine in the molecule provides a more stable anion, because the chlorines are very electronegative and thus withdraw the electrons, which stabilizes the anion. This increase in stability of the anion results in more ionization of the carboxylic acid.

15.83 The hydroxyl group on one end of the molecule can react with the carboxylic acid group of another molecule to create a larger molecule. When this process is repeated a long polymer chain results.

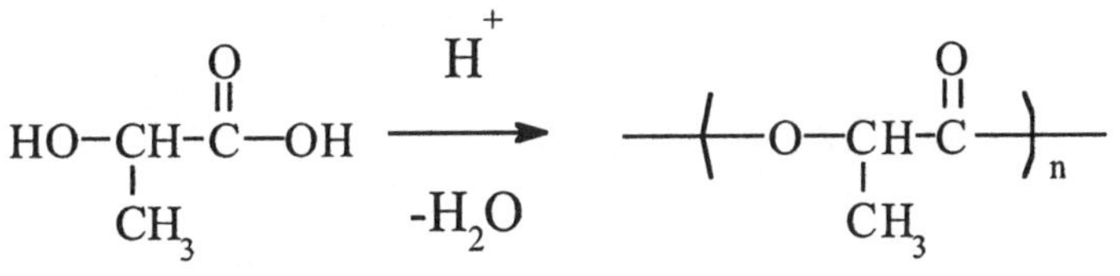

$$\underset{\underset{\displaystyle CH_3}{|}}{HO-CH}-\overset{\overset{\displaystyle O}{\|}}{C}-OH \quad \xrightarrow[-H_2O]{H^+} \quad -\left(\!\!-O-\underset{\underset{\displaystyle CH_3}{|}}{CH}-\overset{\overset{\displaystyle O}{\|}}{C}-\!\!\right)_{\!n}-$$

15.85 The acid is a catalyst, because it increases the rate of the hydrolysis and is not used up. The base, however, is a required reactant that is used up during the hydrolysis reaction.

15.87   This is not an easy analysis because both compounds contain the same functional groups, such as hydroxyls and carboxylic acids. Aspirin does have two carbonyl groups, as it contains both an ester and a carboxylic acid; thus the presence of two carbonyl peaks would identify aspirin. However, the absence of two carbonyl peaks would not preclude aspirin, as the two peaks may be masked into one. The best course of action would be to compare the spectra of a known ibuprofen sample and the spectra of a known aspirin sample to the spectra of the unknown. The known that overlays the unknown is the correct answer, almost no two compounds have exactly the same spectra.

Chapter 16

Amines and Amides

16.1 A tertiary amine means that the nitrogen is bonded to three carbons. Thus, the isomers of $C_5H_{13}N$ must be defined by where the other two carbons that are part of this compound are bonded to the NC_3 center. There are three possible combinations given as:

$$CH_3CH_2—N(CH_3)—CH_2CH_3 \qquad CH_3CH_2CH_2—N(CH_3)—CH_3 \qquad (CH_3)_2CH—N(CH_3)—CH_3$$

16.3 An amide is a compound with a nitrogen that is bonded to a carbonyl group, while an amine is a compound with a nitrogen that is bonded to carbons or hydrogens, but not a carbonyl. A primary amine is one that is bonded to one carbon and two hydrogens, a secondary amine is one that is bonded to two carbons and one hydrogens, and finally, a tertiary amine is one that is bonded to three carbons.

(1) This compound is an amine. Since the N is bonded to two carbons, it is a secondary amine. Because the nitrogen is part of a non-aromatic carbon chain, it is an aliphatic amine.

(2) This compound is an amine. Since the N is bonded to one carbon, it is a primary amine. Because the nitrogen is part of a non-aromatic carbon ring, it is an aliphatic amine. There is also a ketone group in this compound.

(3) This compound is an amine. Since each N has three bonds to carbons, it is a tertiary amine. Because each nitrogen is part of an aromatic carbon ring, it is an aromatic amine.

(4) This compound is an amide because the nitrogen is bonded to a carbonyl.

(5) This compound is an amide because the nitrogen is bonded to a carbonyl.

(6) This is neither an amide nor amine because the nitrogen is bonded to an oxygen as well as a benzene ring, thus it is a nitrobenzene.

16.5 (a) N-methyl-1-pentaneamine: 1-pentaneamine denotes that there is a nitrogen bonded to C-1 of a five carbon (pentane) chain. N-methyl denotes that there is also a methyl group bonded to the nitrogen of the amine, thus:

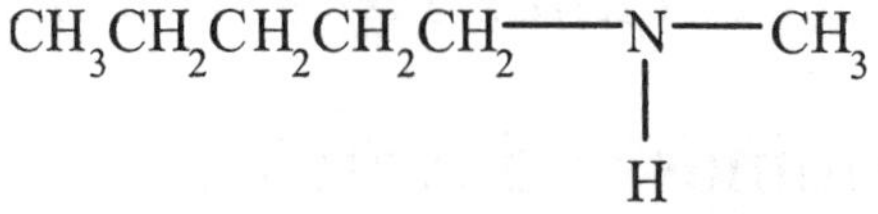

$$CH_3CH_2CH_2CH_2CH_2\!-\!\!\overset{\displaystyle |}{\underset{\displaystyle H}{N}}\!-\!CH_3$$

(b) Isopropylmethylamine: methylamine denotes that there is a methyl group bonded to the nitrogen of an amine, while the isopropyl in the name denotes that there is also an isopropyl group bonded to the nitrogen, thus:

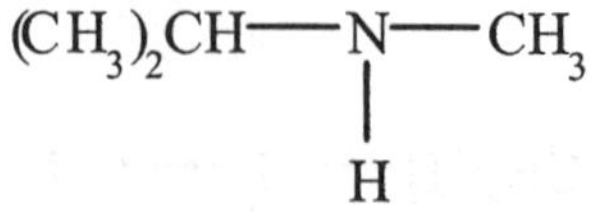

$$(CH_3)_2CH\!-\!\!\overset{\displaystyle |}{\underset{\displaystyle H}{N}}\!-\!CH_3$$

(c) cis-4-ethyl-N-propylcyclohexaneamine: cyclohexanamine denotes that a cyclohexane ring is bonded to the nitrogen of an amine. N-propyl signifies that there is also a propyl group bonded to the nitrogen of the amine. Cis-4-ethyl denotes that there is also an ethyl group bonded to C-4 of the cyclohexane ring (C-1 is the carbon that is bonded to the N) and it is situated on the same side of the ring from the amine, thus:

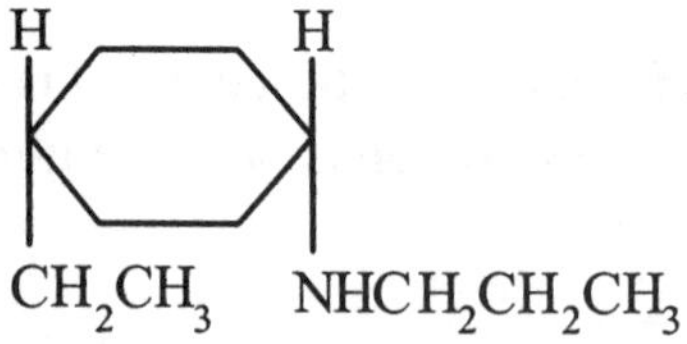

$$CH_2CH_3 \qquad NHCH_2CH_2CH_3$$

(d) N-ethylaniline: aniline is benzene with an NH_2 group connected to it. The N-ethyl denotes that there is also an ethyl group bonded to the nitrogen of the amine, thus:

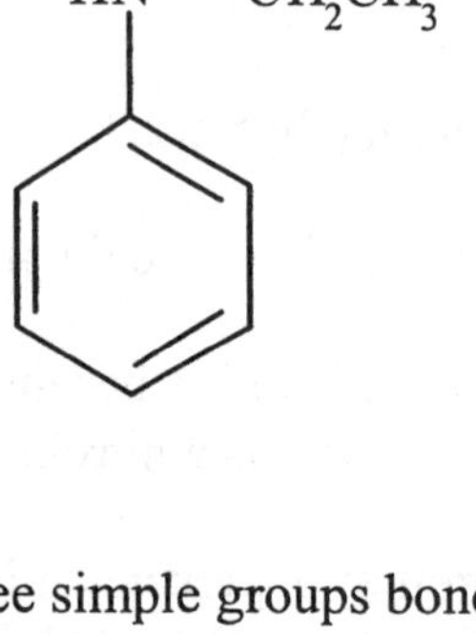

$$HN\!-\!CH_2CH_3$$

16.7   (a) This is a tertiary amine with three simple groups bonded to the nitrogen. These groups are two methyl groups and an ethyl group. Putting the group names in front of the word "amine" completes naming the amine, thus this compound is ethyldimethylamine.

(b) This compound has a cyclohexane ring bonded to the amine group, thus the base name is cyclohexanamine. There is also a methyl group bonded to the cyclohexane ring at C-2, trans to the amine, thus this compound is trans-2-methylcyclohexanamine.

(c) The groups bonded to the nitrogen are not simply named, thus the method used in (a) will not work for this compound. Thus the longest chain that has a carbon to which the nitrogen is bonded is found. That is a five-carbon chain, pentane. On the pentane chain is a methyl group at C-4 and the amine at C-2. Thus, the base name is 4-methyl-2-pentanamine. The compound is slightly more complex as it has a methyl and an ethyl group on the nitrogen. This is denoted in the name by putting N-methyl and N-ethyl in front of the base name. Because there are two methyl groups, they are combined to form the name N-ethyl-N, 4-dimethyl-2-pentanamine.

(d) The nitrogen bonded to the benzene ring is aniline. The ethyl and methyl groups that are also bonded to the nitrogen are denoted by putting N-ethyl and N-methyl in front of aniline, thus this compound is N-ethyl-N-methylaniline.

16.9   An amine can form a hydrogen bond between one of the amino hydrogens and the nitrogen of another molecule as:

$$C_2H_5-N\begin{matrix} H \\ \\ H \end{matrix} \quad \cdots \cdots \quad H-N-C_2H_5$$

16.11   (a) The O–H bond in an alcohol is more polar than the N–H bond of an amine. This increased polarity results in stronger hydrogen bonds between two alcohol molecules than the hydrogen bonds that exist between two amine molecules. This stronger secondary attractive interaction results in 1-butanol having a higher boiling point than butylamine.

(b) The hydrogen bond between water and the N–H of an amine is similar to the hydrogen bond between water and the carbonyl oxygen of an aldehyde. This results in the solubilities of butylamine and butanal being about the same.

(c) The primary amine butylamine can form hydrogen bonds between two molecules because there is a hydrogen bonded to the nitrogen. The tertiary amine ethyldimethylamine, however, cannot form hydrogen bonds. Therefore, the secondary attractive forces between butylamine molecules are greater than those of ethyldimethylamine that results in butylamine having a higher boiling point than ethyldimethylamine.

(d) Both primary and secondary amines can hydrogen bond with water, thus they have similar solubilities in water.

16.13   A base can be defined as a proton acceptor. Therefore, propylamine must take the hydrogen from HCl to form a salt as:

$$CH_3CH_2CH_2NH_2 + HCl \rightarrow CH_3CH_2CH_2NH_3^+ \ Cl^-$$

16.15   Basicity measures the ability of a compound to abstract a proton. Amines are more basic than water or an alcohol because the nitrogen of the amine has two non-bonded electrons that can bond to a proton. Aniline is less basic than propylamine because the electron withdrawing effect of the aromatic ring makes the non-bonded electrons of the nitrogen less available. KOH is the strongest base because the base is actually OH⁻ which seeks the positive charge of the proton to neutralize the anion. Thus,

$$\text{water} \approx \text{propanol} < \text{aniline} < \text{propylamine} < \text{KOH}$$

16.17   A compound that turns red litmus paper blue is a base, thus the unknown must be propylamine.

16.19   A base can be defined as a proton acceptor and each NH$_2$ group will act as a base toward HCl. Therefore, 1,6-hexanediamine must take the hydrogen from HCl to from a salt as:

$$H_2NCH_2CH_2CH_2CH_2CH_2CH_2NH_2 + 2 \ HCl \rightarrow Cl^- \ H_3NCH_2CH_2CH_2CH_2CH_2CH_2NH_3^+ \ Cl^-$$

16.21   (a) N, N-dimethylcyclopentanammonium chloride: ammonium chloride denotes that this is an ammonium salt and the counter-ion is chloride. Cyclopentanammonium signifies that the base structure of the amine from which the salt was derived had an amine bonded to a cyclopentane ring. Finally, N, N-dimethyl denotes that there are two methyl groups that are bonded to the nitrogen. Thus the structure is:

(b) N-ethyl-N-methylpiperidinium sulfate: piperidinium sulfate denotes that this is an ammonium salt that is derived from piperidine (the six-membered ring with a nitrogen in the ring) and the counter-ion is chloride. The N-ethyl and N-methyl denote that there is also a methyl group and an ethyl group that is bonded to the nitrogen. Thus the compound is:

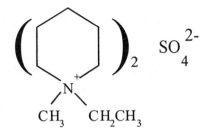

(c) Ethylmethylpropylammonium bromide: ammonium chloride denotes that this is an ammonium salt and the counter-ion is bromide. The "ethylmethylpropyl" before the "ammonium" in the name denotes that the amine from which this salt was derived had an ethyl, a methyl, and a propyl group bonded to the nitrogen, thus the compound is:

$$\left[CH_3CH_2\!\!-\!\!\overset{\displaystyle CH_3}{\underset{\displaystyle CH_2CH_2CH_3}{\overset{|}{\underset{|}{\overset{+}{N}}}}}\!\!-\!\!H \right] \quad Br^-$$

16.23 (a) An amine salt is an ionic compound. Ionic interactions are much stronger than the hydrogen bonding secondary forces of an amine, thus trimethylammonium chloride has a higher boiling point than triethylamine.

(b) Amine salts can completely dissociate into ions that interact more strongly with water than an amine can by hydrogen bonding. Thus, hexylammonium chloride has a higher solubility in water than hexylamine.

16.25 An amine is a base that will change red litmus paper blue, while an ammonium salt is an acid that will turn blue litmus paper red, thus this compound is an amine, butylamine.

16.27 An amide is a compound that has a nitrogen that is bonded to a carbonyl group. Amides are classified as primary when two hydrogens are bonded to the nitrogen, as secondary when one hydrogen is bonded to nitrogen, and as tertiary when no hydrogens are bonded to the nitrogen.

(1) This compound does have a nitrogen bonded to a carbonyl, thus it is an amide. There are no hydrogens on the nitrogen, thus it is a tertiary amide.

(2) This compound does not have a nitrogen bonded to a carbonyl, thus it is not an amide, it is a ketone and an amine.

(3) This compound does have a nitrogen bonded to a carbonyl, thus it is an amide. There are two hydrogens on the nitrogen, thus it is a primary amide. There also exists an

NH$_2$ group at the other end of the molecule, which is a primary amine.

(4) This compound does have a nitrogen bonded to a carbonyl, thus it is an amide. There is one hydrogen on the nitrogen, thus it is a secondary amide.

16.29 (a) A carboxylic acid will react with an amine at room temperature to undergo an acid-base reaction to form an ammonium salt:

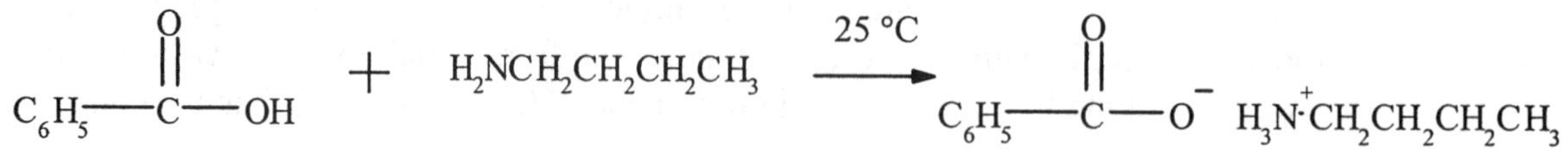

$$C_6H_5\text{---}\overset{\displaystyle O}{\overset{\|}{C}}\text{---}OH \quad + \quad H_2NCH_2CH_2CH_2CH_3 \quad \xrightarrow{25\,°C} \quad C_6H_5\text{---}\overset{\displaystyle O}{\overset{\|}{C}}\text{---}O^- \;\; H_3\overset{+}{N}CH_2CH_2CH_2CH_3$$

(b) A carboxylic acid and amine will undergo a coupling reaction to form an amide at elevated temperature, greater than 100°C.

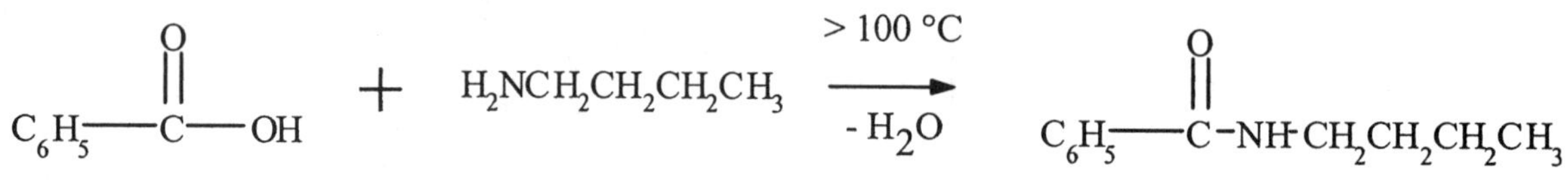

$$C_6H_5\text{---}\overset{\displaystyle O}{\overset{\|}{C}}\text{---}OH \quad + \quad H_2NCH_2CH_2CH_2CH_3 \quad \xrightarrow[-H_2O]{>\,100\,°C} \quad C_6H_5\text{---}\overset{\displaystyle O}{\overset{\|}{C}}\text{-NH-}CH_2CH_2CH_2CH_3$$

(c) An acid chloride is more reactive than a carboxylic acid, therefore an acid chloride will react with an amine at room temperature to form an amide.

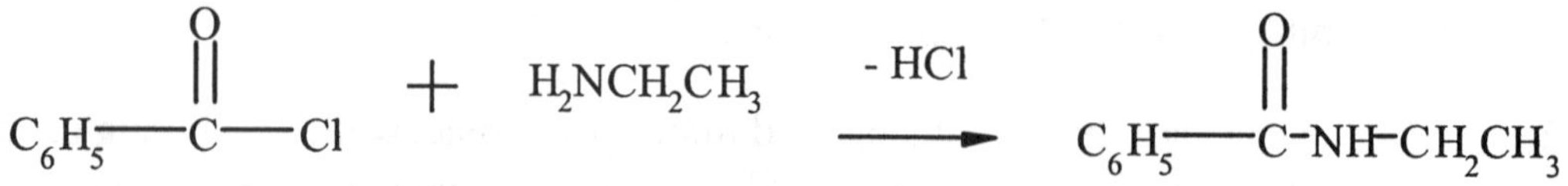

$$C_6H_5\text{---}\overset{\displaystyle O}{\overset{\|}{C}}\text{---}Cl \quad + \quad H_2NCH_2CH_3 \quad \xrightarrow{-\,HCl} \quad C_6H_5\text{---}\overset{\displaystyle O}{\overset{\|}{C}}\text{-NH-}CH_2CH_3$$

(d) An anhydride will react with an amine to form an amide and a carboxylic acid.

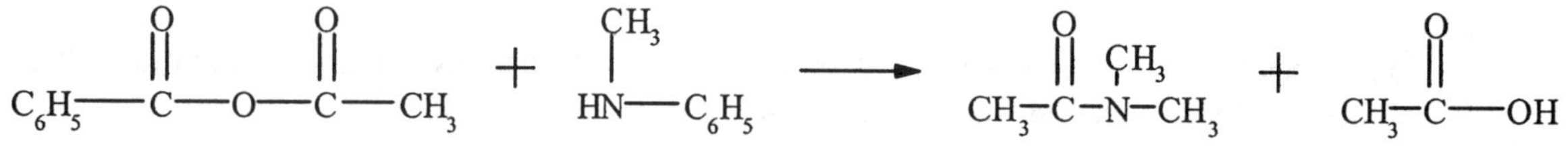

$$C_6H_5\text{---}\overset{\displaystyle O}{\overset{\|}{C}}\text{---}O\text{---}\overset{\displaystyle O}{\overset{\|}{C}}\text{---}CH_3 \quad + \quad \overset{\displaystyle CH_3}{\underset{HN\text{---}C_6H_5}{|}} \quad \longrightarrow \quad CH_3\text{-}\overset{\displaystyle O}{\overset{\|}{C}}\text{-}\overset{CH_3}{\underset{\;}{N}}\text{-}CH_3 \quad + \quad CH_3\text{---}\overset{\displaystyle O}{\overset{\|}{C}}\text{---}OH$$

16.31 To determine what carboxylic acid and amine are needed to form an amide, merely break the nitrogen–carbonyl bond in the amide, add an –OH to the carbonyl to form the carboxylic acid, and add a hydrogen to the nitrogen to form the amine, thus:

(a)

$$(CH_3)_3CCH_2CH_2\underset{\displaystyle \overset{\displaystyle O}{\|}}{C}-OH \quad + \quad H_2NCH(CH_3)_2$$

(b)

$$2\ CH_3CH_2\underset{\displaystyle \overset{\displaystyle O}{\|}}{C}-OH \quad + \quad H_2NCH_2CH_2CH_2NH_2$$

16.33 To determine what dicarboxylic acid and diamine are needed to form a polyamide, merely break the nitrogen–carbonyl bond in the two amide linkages of the polyamide. Then, add an –OH to both carbonyls to form the dicarboxylic acid and add a hydrogen to both nitrogens to form the diamine, thus:

$$HO-\overset{\displaystyle \overset{\displaystyle O}{\|}}{C}-CH_2CH_2CH_2CH_2-\overset{\displaystyle \overset{\displaystyle O}{\|}}{C}-OH \quad + \quad H_2NCH_2CH_2CH_2NH_2$$

16.35 (a) N-*t*-butylbutanamide: butanamide means that the amide is derived from butanoic acid, thus there are three carbons bonded to the carbonyl. N-*t*-butyl denotes that there is also a tert-butyl group bonded to the nitrogen of the amide, thus the structure is:

$$CH_3CH_2CH_2-\overset{\displaystyle \overset{\displaystyle O}{\|}}{C}-NHC(CH_3)_3$$

(b) N-phenyl-3-methylhexanamide: hexanamide means that the amide is derived from hexanoic acid, which means that there are five carbons chain bonded to the carbonyl. The 3-methyl denotes that there is a methyl group on C-3 (where C-1 is the carbonyl carbon). Finally, N-phenyl denotes that there is also a phenyl group bonded to the amide nitrogen, thus the structure is:

$$\underset{CH_3CH_2CH_2}{}\!-\!\underset{\overset{|}{CH_3}}{CH}\!-\!CH_2\!-\!\underset{\overset{\|}{O}}{C}\!-\!NHC_6H_5$$

16.37 The first step in naming amides should be the determination of the carboxylic acid from which the amide derives. Replacing the carbonyl–nitrogen bond with a carboxylic acid group does this. The base name of the amide is formed from this carboxylic acid by replacing "oic acid" with "amide". Placing the name of the group preceded by "N" in front of the base name then specifies groups on the nitrogen.

(a) The carboxylic acid from which this derives is pentanoic acid, thus the base name is pentanamide. There is also an isopropyl group bonded to the nitrogen, which is denoted by placing N-isopropyl in front of pentanamide, thus this compound is N-isopropylpentanamide.

(b) The carboxylic acid from which this derives is propanoic acid, thus the base name is propanamide. There is also a phenyl group bonded to the nitrogen, which is denoted by placing N-phenyl in front of propanamide, thus this compound is N-phenylpropanamide.

(c) The carboxylic acid from which this derives is benzoic acid, thus the base name is benzamide. There are also two ethyl groups bonded to the nitrogen, which is denoted by placing N, N-dimethyl in front of benzamide, thus this compound is N, N-dimethylbenzamide.

16.39 (a) Pentanamide has a hydrogen connected to the nitrogen, which can undergo hydrogen bonding to the carbonyl of another molecule. N, N-dimethylpropanamide does not have a hydrogen connected to the nitrogen and thus cannot form hydrogen bonds between molecules. Thus pentanamide will have a higher boiling point than N, N-dimethylpropanamide.

(b) Both pentanamide and N, N-dimethylpropanamide can form hydrogen bonds to water through the carbonyl oxygen. Pentanamide can also hydrogen bond to water through the N–H, and thus the two compounds have very similar water solubilities, but pentanamide is slightly more soluble due to the increased susceptibility to forming hydrogen bonds.

(c) The hydrogen bond between two amide molecules is stronger than the hydrogen bond between two carboxylic acid molecules, thus butanamide has a higher boiling point than butanoic acid.

(d) Both compounds can hydrogen bond with water, but the hydrogen bond involving the amide group is slightly stronger than the hydrogen bond involving a carboxylic acid.

Thus, the compounds will have similar solubilities, but hexanamide will be slightly more soluble.

(e) Both compounds can hydrogen bond with water, but the hydrogen bond involving the amide group is slightly stronger than those involving an alcohol. Thus, butanamide will be more soluble.

16.41 Basicity measures the ability of a compound to abstract a proton. Amines are more basic than water or an alcohol because the nitrogen of the amine has two non-bonded electrons that can bond to a proton. Amides are no more basic than alcohols or water because the proximity of the carbonyl to the nitrogen limits the availability of the non-bonded electrons of the nitrogen. NaOH is the strongest base because the base is actually OH⁻ which seeks the positive charge of the proton to neutralize the anion. Thus,

water ≈ 1-propanol ≈ ethanamide < propylamine < NaOH

16.43 (a) In the presence of an acid and water, an amide readily hydrolyzes (adds water) at the O=C–N bond to form a carboxylic acid and an amine salt.

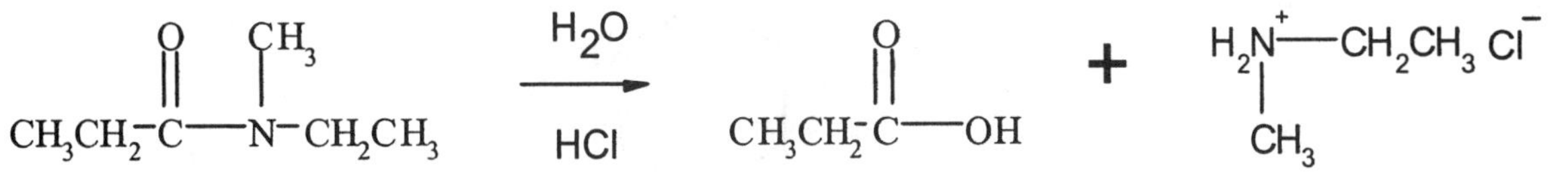

(b) In the presence of a base and water, an amide readily hydrolyzes (adds water) at the O=C–N bond to form a carboxylate salt and an amine.

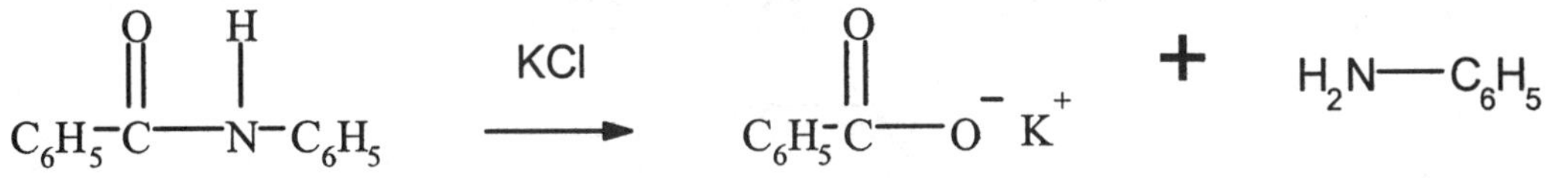

16.45 (a) Amides are classified as primary when two hydrogens are bonded to the nitrogen. Thus the three carbons must be part of the carbonyl as well as the hydrocarbon group bonded to the carbonyl. Thus the structure is:

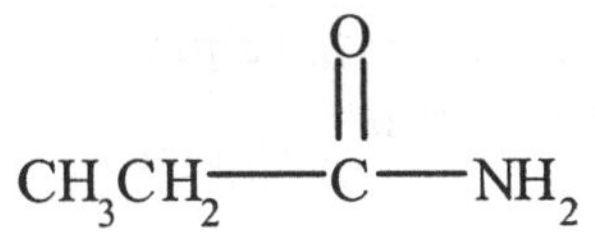

(b) A tertiary amide has no hydrogens bonded to the nitrogen. The fact that the compound has three methyl groups means that there must be a methyl group bonded to the carbonyl and two bonded to the nitrogen. This gives the four carbons the problem, thus the structure is:

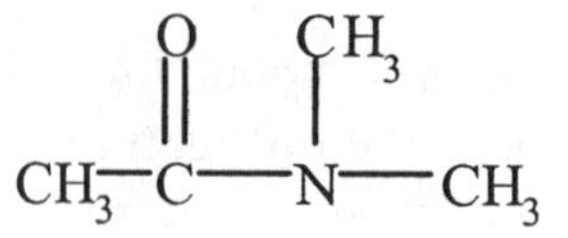

$$CH_3-C(=O)-N(CH_3)-CH_3$$

(c) A lactam is a ring structure with an amide group in the ring. The question also states that the structure has five carbons that are all in the ring, thus:

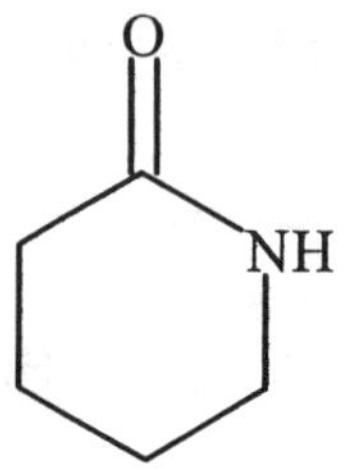

(d) A secondary amine is one that has a nitrogen that is bonded to one hydrogen and two carbons. One of the groups that the nitrogen is bonded to must be a benzene rig, C_6H_5. Thus that leaves the CH_3 group to be bonded to the nitrogen ($C_7H_9N - C_6H_5$ [of benzene ring] $-$ NH [of amine] $= CH_3$). Thus the structure must be:

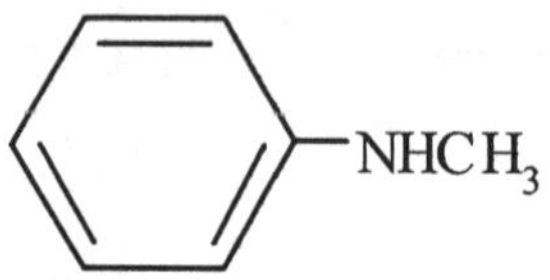

(e) A tertiary amine is one that has a nitrogen that is bonded to three carbons. The *t*-butyl group accounts for C_4H_9 ($= C(CH_3)_3$) of the hydrocarbon structures that are bonded to the nitrogen. Thus there remains C_2H_6 to be distributed as two hydrocarbon groups that are bonded to the nitrogen so that the amine is tertiary. This can only be done as two $-CH_3$ groups, thus the structure is:

$$CH_3-N-C(CH_3)_3$$
$$|$$
$$CH_3$$

16.47 (a) 5-amino-3-methyl-2-pentanone: 2-pentanone denotes that this compound is a ketone with a five-carbon chain and the carbonyl at C-2. 3-methyl means that there is also a methyl group on C-3 while 5-amino denotes that there is an amino (NH_2) group on C-5. This is not an amide because the nitrogen is not bonded to the carbonyl.

$$\overset{O}{\overset{||}{CH_3-C}}-\overset{\overset{CH_3}{|}}{CH}-CH_2CH_2NH_2$$

(b) 4-pentene-1-amine: 4-pentene denotes that the compound has a five-carbon chain with a double bond at C-4. 1-amine at the end also denotes that there is an amino group bonded to C-1, thus:

$$H_2NCH_2CH_2CH_2CH=CH_2$$

16.49 (a) Atoms a, b, and c do not involve double bonds and therefore they are all sp^3 hybridized. Angles A and B only involve C–C and C–N bonds and are thus 109.5°.

(b) Atoms a, b, and c do not involve double bonds and therefore they are all sp^3 hybridized. Angles A and B only involve C–C, N–H, and C–N bonds and are thus 109.5°.

16.51 (a) The hydrogen bond between two amide molecules is stronger than the hydrogen bond between two amine molecules, thus butanamide has a higher boiling point than hexanamine.

(b) Both compounds can hydrogen bond with water, but the hydrogen bond involving the amide group is slightly stronger than those involving an amine. Thus, the compounds will have similar solubilities, but butanamide will be slightly more soluble.

(c) The hydrogen bond between two amide molecules is stronger than the hydrogen bond between two carboxylic acid molecules, thus pentanamide has a higher boiling point than pentanoic acid.

(d) Both compounds can hydrogen bond with water, but the hydrogen bond involving the amide group is slightly stronger than those involving a carboxylic acid. Thus, the compounds will have similar solubilities, but pentanamide will be slightly more soluble.

(e) An amine salt is an ionic compound. Ionic interactions are much stronger than the hydrogen bonding secondary forces of an amide, thus trimethylammonium chloride has a higher boiling point than hexanamide.

(f) A carboxylate salt is an ionic compound. Ionic interactions are much stronger than the hydrogen bonding secondary forces of an amide, thus sodium butanoate has a higher boiling point than pentanamide.

(g) The two compounds both have a single NH_2 group for every four carbons of the compound. The NH_2 groups can hydrogen bond to water, thus the two compounds have the same solubility in water.

16.53   An amine is a base that will turn red litmus paper blue, while an amide will not. Thus, add a drop of the unknown to red litmus paper. If it turns blue it is an amine, if it does not, it is an amide.

16.55   An amine is a base that will turn red litmus paper blue, while an amide or an alcohol will not. Therefore, the unknown must be either propanol or ethanamide.

16.57   A polymer only forms if both reactants are bifunctional. In this problem, only (c) has two bifunctional compounds that form a polymer as:

16.59   (a) The first compound is an amide with five carbons, while the second is an amine and ketone with six carbons. In other words in the first compound the nitrogen is bonded to the carbonyl, while in the second it is not, therefore, these are different compounds that are not isomers.

(b) Both compounds have the same molecular formula $C_5H_{11}NO$, but compound 1 is an amide, while the second is an amine and ketone, thus these two compounds are different compounds that are constitutional isomers.

(c) Both compounds have the same molecular formula C_3H_9N, but compound 1 is a tertiary amine, while the second is a primary amine, thus these two compounds are different compounds that are constitutional isomers.

(d) Both compounds have the same molecular formula C_5H_9NO, but compound 1 is a five-membered heterocyclic amine, while the second is a six-membered heterocyclic amine, thus these two compounds are different compounds that are constitutional isomers.

16.61 $K_b = \dfrac{[OH^-][CH_3NH_3^+]}{[CH_3NH_2]} = 4.59 \times 10^{-4}$

$[OH^-] = [CH_3NH_3^+]$ $[CH_3NH_2] = 1.00$

$$[OH^-] = (4.59 \times 10^{-4})^{0.5} = 2.14 \times 10^{-2} \qquad\qquad [H_3O^+][OH^-] = 1.0 \times 10^{-14}$$

$$[H_3O^+] = 1.0 \times 10^{-14} / [OH^-] = 1.0 \times 10^{-14} / 2.14 \times 10^{-2} = 4.6 \times 10^{-13}$$

$$pH = -\log[H_3O^+] = 12.33$$

16.63 An amide will undergo hydrolysis to form a carboxylic acid and an amine. If the amine that is formed is ethylamine ($NH_2CH_2CH_3$), then the carboxylic acid must have three carbons in it (C_5 from amide – C_2 from amine). Therefore, the other compound is a carboxylic acid with three carbons, propanoic acid.

16.65 Ammonium chloride is an ionic compound that is not volatile. This lack of volatility results in a compound that does not have a strong odor.

16.67 Methylamine by itself will react with water to form methyl ammonium ion. The presence of HCl will convert all methylamine to methyl ammonium, while the presence of NaOH will suppress the reaction of methylamine with water to form methyl ammonium. Thus, B > A > C.

16.69 The physical strength of a polymer increases with increasing secondary forces between molecules, just as boiling and melting points. Nylon-66 molecules can undergo intermolecular hydrogen bonding between the carbonyl and N–H groups, which are much greater than the London forces between polyethylene molecules.

16.71 A soap must have both an ionic end group and a long hydrocarbon chain. Only compound 2 has both of these, thus it is the only soap.

16.73 The free amine is a covalent compound with relatively weak secondary forces and volatilizes easily at moderated temperatures. The amine salt is an ionic compound with very strong secondary forces and will not volatilize at moderate or even high temperatures.

16.75 Phenobarbital contains amide groups, which include carbonyl groups that have peaks in the IR spectrum at 1630–1690 cm^{-1}. Methamphetamine does not possess carbonyl groups and thus would not show any peaks in the IR spectrum in this region.

16.77 $(CH_3)_3N + CH_3Cl \rightarrow (CH_3)_4N^+Cl^-$

16.79 Counting the number of each atom present: $C_8H_{10}N_4O_2$

16.81 Because Y ($C_2H_4O_2$) reacts with ethylamine (C_2H_7N) in the presence of an acid to yield C_4H_9NO, this is clearly an addition reaction with water given off, thus $C_2H_4O_2$ must be a carboxylic acid and C_4H_9NO must be an amide. As aldehydes are oxidized to carboxylic acids by adding one oxygen ($C_2H_4O \rightarrow C_2H_4O_2$), X must be an aldehyde. Thus, X is ethanal, Y is ethanoic acid, and Z = N-ethylethanamide.

Chapter 17

Stereoisomerism

17.1 Constitutional isomers are compounds that have the same molecular formula, but different connectivity. Thus, isomers of C_4H_{10} will be different compounds that differ in the way that the carbon atoms are bonded together. The most obvious first choice is to have them all bonded in a straight chain as $CH_3CH_2CH_2CH_3$. Next, the chain can be branched. The only possible branch point is on C3, thus moving the C1 methyl group to be a branch at C3 gives $CH_3C(CH_3)CH_3$. These are the only two constitutional isomers of C_4H_{10}.

17.3 Butanol has the molecular formula $C_4H_{10}O$. An ether is a compound with formula R1–O–R2, where R1, and R2 are hydrocarbon structures. Possible combinations of R1 and R2 that can be formed from C_4H_{10} are R1 = C_2H_5 and R2 = C_2H_5; R1 = CH_3 and R2 = C_3H_7 (linear); R1 = CH_3 and R2 = C_3H_7 (branched); thus the isomers are $CH_3CH_2OCH_2CH_3$, $CH_3OCH_2CH_2CH_3$, and $CH_3OCH(CH_3)_2$.

17.5 *cis*-2-butene is a compound with a double bond. The diastereomer of a compound with a double bond is a *cis-trans* isomer, thus the diastereomer of *cis*-2-butene will be *trans*-2-butene:

17.7 A chiral object is one that cannot be superimposed on its mirror image:

(a) A person cannot be superimposed on its mirror image, thus it is chiral.

(b) An automobile cannot be superimposed on its mirror image, thus it is chiral.

(c) A basketball with no writing can be superimposed on its mirror image, thus it is achiral.

(d) The addition of a written name on a basketball results in an object that cannot be superimposed on its mirror image, thus it is chiral.

17.9 For a compound to be able to form enantiomers, it must contain a carbon atom that is bonded to four different structures.

(a) Every carbon has at least two hydrogens except C3. C3 is bonded to an –H, to –CH₃, and two $-C_2H_5$ groups. Therefore, there is no carbon in this molecule that is bonded to four different groups, thus it cannot exist as enantiomers.

(b) Every carbon has at least two hydrogens except C2. C2 is bonded to an $-H$, to $-C_3H_7$, and two $-CH_3$ groups. Therefore, there is no carbon in this molecule that is bonded to four different groups, thus it cannot exist as enantiomers.

(c) Every carbon has at least two hydrogens except C2. C2 is bonded to four different groups, $-CH_2OH$, $-CH_3$, $-H$, and $-NH_2$. Thus this compound can exist as two enantiomers, which are:

$$\begin{array}{ccc} CH_2OH & \quad & CH_2OH \\ | & & | \\ H\!-\!\!C^*\!\!-\!NH_2 & & H_2N\!-\!\!C^*\!\!-\!H \\ | & & | \\ CH_3 & & CH_3 \end{array}$$

(d) Every carbon is either part of a double bond (aromatic ring) or has at least two hydrogens. Therefore, there is no carbon in this molecule that is bonded to four different groups, thus it cannot exist as enantiomers.

(e) Every carbon is either part of a double bond (aromatic ring) or has at least two hydrogens except the carbon that is connected to the phenyl ring. This carbon is bonded to four different groups, $-C_6H_5$, $-CH_3$, $-H$, and $-NHCH_3$. Thus this compound can exist as enantiomers, which are:

$$\begin{array}{ccc} CH_2OH & \quad & CH_2OH \\ | & & | \\ H\!-\!\!C^*\!\!-\!NH_2 & & H_2N\!-\!\!C^*\!\!-\!H \\ | & & | \\ CH_3 & & CH_3 \end{array}$$

17.11 When comparing two compounds, the first action should be to count the number of each type of atom to find a molecular formula. If they are not the same, then the two compounds must be different compounds that are not isomers. Inspection of the compounds in this question shows that this is not true for any of the molecules in this question.

The next step should be to determine if the two molecules are the same molecule. In this question, that can best be determined by superimposing the two molecules on top of each other. Care must be taken here to preserve the Fischer projection. In (b), if either molecule is rotated 180° in the plane of the page, the other molecule will be formed, i.e., the two molecules become superimposable. In (c), the two molecules are superimposable as written. In (e) the two molecules are superimposable as written.

Further inspection differentiates between enantiomers, diastereomers, or constitutional isomers. In constitutional isomers, the connectivity of the atoms is different while enantiomers or diastereomers have the same connectivity, yet the two molecules cannot be superimposed. Inspection shows that none of the compounds has different connectivity. Finally, for two molecules to be enantiomers, they must be mirror images of each other and have the same connectivity. In (a) and (d), it can be seen that the two molecules are mirror images if the first compound is placed above the second.

Thus:

(a) 3, (b) 1, (c) 1, (d) 3, (e) 1

17.13   To identify D- and L-enantiomers, the compound must be written in the proper manner. The –H and –X groups must be in the horizontal and the two hydrocarbon substituents must be in the vertical plane on a Fischer projection. Moreover, R_1, the hydrocarbon with the fewest hydrogens, must be at the top of the figure and R_2, the compound with the most hydrogens, must be at the bottom. When the compound is written this way, a D-enantiomer has the hydrogen on the left, and the L-enantiomer has it on the right.

(a) This Fischer projection is written correctly, thus it can be used to determine the D or L designation. This projection has the –H on the left side and is thus the D-enantiomer.

(b) This projection does not fit the rules given above, thus it must be rotated 180° in the plane of the page to give the aldehyde group on top and the alcohol group on the bottom. When the compound is drawn this way, the –H is on the right, thus it is the L-enantiomer.

(c) This Fischer projection is written correctly, thus it can be used to determine the D or L designation. This projection has the –H on the left side and is thus the D-enantiomer.

(d) This projection does not fit the rules given above, thus it must be rotated 180° in the plane of the page to put the carboxylic acid group on top and the methyl group on the bottom. When the compound is drawn this way, the –H is on the right, thus it is the L-enantiomer.

17.15   (a) Ethanol is not chiral and therefore will not rotate the plane of polarized light.

(b) D-glucose is chiral and therefore will rotate the plane of polarized light. However, the D configuration does not specify which direction it will rotate the plane. It must be experimentally determined.

(c) (+)-phenylalanine is chiral and will rotate the plane of polarized light. The (+) denotes that it will rotate in the clockwise direction.

(d) A racemic mixture of glutamic acid is not chiral and will not rotate the plane of polarized light.

17.17   The D-enantiomer of a compound will rotate the plane of polarized light in the opposite direction and to the same extent as the L-enantiomer, thus the specific rotation of L-glutamic acid is +31.5°.

17.19 The (+) enantiomer of alanine must be the mirror image of (–)-alanine:

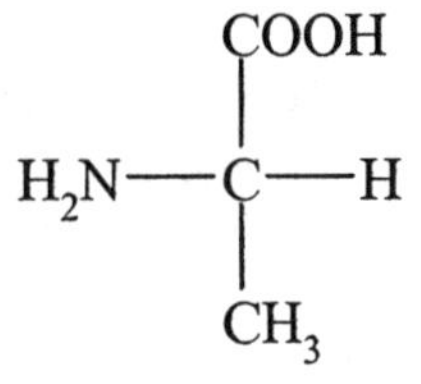

$$
\begin{array}{c}
COOH \\
| \\
H_2N-C-H \\
| \\
CH_3
\end{array}
$$

17.21 The only chemical properties that chirality affects are optical rotation and chiral recognition. Thus, the water solubility of D-alanine at 25°C is equal to the water solubility of L-alanine, 127 g/L.

17.23 The specific rotation [α] can be determined using the equation $[\alpha]=\dfrac{\alpha}{CL}$ where α is the observed rotation in degrees for a solution with concentration C (in g/mL) through a cell of length L (in dm = 10 cm). α = –3.6°, C = 0.040 g/mL, L = 1.00 dm, thus [α] = –90°.

17.25 $[\alpha]=\dfrac{\alpha}{CL}$ → [α] = +223°, C = 0.0600 g/ml, L = 1.50 dm. Thus α = [α]CL = +20.1°.

17.27 (a) In this compound there is one carbon atom that is bonded to four different groups, which gives the following pair of enantiomers, which are both optically active.

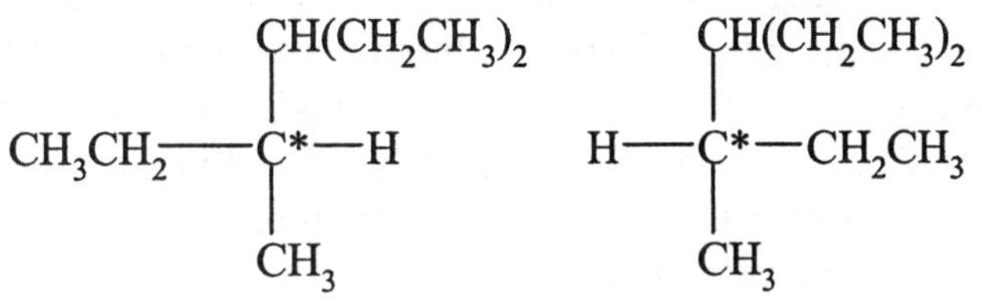

$$
\begin{array}{cc}
CH(CH_2CH_3)_2 & CH(CH_2CH_3)_2 \\
| & | \\
CH_3CH_2-C{*}-H \qquad\qquad & H-C{*}-CH_2CH_3 \\
| & | \\
CH_3 & CH_3
\end{array}
$$

(b) There exist two carbons that are bonded to four different groups. Therefore, it may form four stereoisomers. However inspection of these structures shows that two of them (3) and (4) are equivalent and therefore this is a meso compound. (1) and (2) are enantiomers and (3) is a meso compound. (1) and (2) are diastereomers of (3) and vice versa. Structures (1) and (2) are optically active but structure (3) is not.

$$
\begin{array}{cccc}
\underset{\text{1}}{
\begin{array}{c}
CH_2CH_3 \\
| \\
H\!-\!C^*\!-\!CH_3 \\
| \\
CH_3\!-\!C^*\!-\!H \\
| \\
CH_2CH_3
\end{array}}
&
\underset{\text{2}}{
\begin{array}{c}
CH_2CH_3 \\
| \\
CH_3\!-\!C^*\!-\!H \\
| \\
H\!-\!C^*\!-\!CH_3 \\
| \\
CH_2CH_3
\end{array}}
&
\underset{\text{3}}{
\begin{array}{c}
CH_2CH_3 \\
| \\
CH_3\!-\!C^*\!-\!H \\
| \\
CH_3\!-\!C^*\!-\!H \\
| \\
CH_2CH_3
\end{array}}
&
=\quad
\underset{\text{4}}{
\begin{array}{c}
CH_2CH_3 \\
| \\
CH_3\!-\!C^*\!-\!H \\
| \\
CH_3\!-\!C^*\!-\!H \\
| \\
CH_2CH_3
\end{array}}
\end{array}
$$

(c) There exist two carbons that are bonded to four different groups. Therefore, it may form four stereoisomers. Structures (1) and (2) are diastereomers of (7) and (8) and vice versa. Structures (1) and (2) are enantiomers as are (3) and (4). All four compounds are optically active.

$$
\begin{array}{cccc}
\underset{\text{1}}{
\begin{array}{c}
CH_2CH_3 \\
| \\
H\!-\!C^*\!-\!OH \\
| \\
CH_3\!-\!C^*\!-\!H \\
| \\
CH_2CH_3
\end{array}}
&
\underset{\text{2}}{
\begin{array}{c}
CH_2CH_3 \\
| \\
HO\!-\!C^*\!-\!H \\
| \\
H\!-\!C^*\!-\!CH_3 \\
| \\
CH_2CH_3
\end{array}}
&
\underset{\text{3}}{
\begin{array}{c}
CH_2CH_3 \\
| \\
HO\!-\!C^*\!-\!H \\
| \\
CH_3\!-\!C^*\!-\!H \\
| \\
CH_2CH_3
\end{array}}
&
\underset{\text{4}}{
\begin{array}{c}
CH_2CH_3 \\
| \\
H\!-\!C^*\!-\!OH \\
| \\
H\!-\!C^*\!-\!CH_3 \\
| \\
CH_2CH_3
\end{array}}
\end{array}
$$

17.29 The difference in the boiling point and optical activity signifies that these two compounds are not enantiomers. Therefore, the other two compounds must be the enantiomers of these two compounds. Structure (3) is the enantiomer of (1); structure (4) is the enantiomer of (2). Thus, the boiling point of (3) will be the same as (1), 180–182 °C, and its specific rotation will be –15°. The boiling point of (4) will be 163–165°C, the same as (2) and its specific rotation will be –26°.

$$
\begin{array}{cc}
\underset{\text{3}}{
\begin{array}{c}
CH_3 \\
| \\
HO\!-\!C^*\!-\!H \\
| \\
H_3C\!-\!C^*\!-\!H \\
| \\
CH_2OH
\end{array}}
&
\underset{\text{4}}{
\begin{array}{c}
CH_3 \\
| \\
HO\!-\!C^*\!-\!H \\
| \\
H\!-\!C^*\!-\!CH_3 \\
| \\
CH_2OH
\end{array}}
\end{array}
$$

17.31 (a) Each carbon that is bonded to four different groups is a tetrahedral stereocenter. There are four such carbons in this molecule:

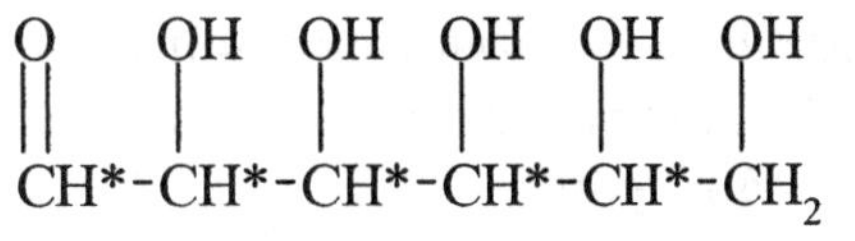

$$\overset{O}{\underset{||}{C}}H^*-\overset{OH}{\underset{|}{C}}H^*-\overset{OH}{\underset{|}{C}}H^*-\overset{OH}{\underset{|}{C}}H^*-\overset{OH}{\underset{|}{C}}H^*-\overset{OH}{\underset{|}{C}}H_2$$

A molecule with n tetrahedral stereocenters can form 2^n stereoisomers, thus this molecule can form 16 stereoisomers.

(b) Each carbon that is bonded to four different groups is a tetrahedral stereocenter. There are three such carbons in this molecule:

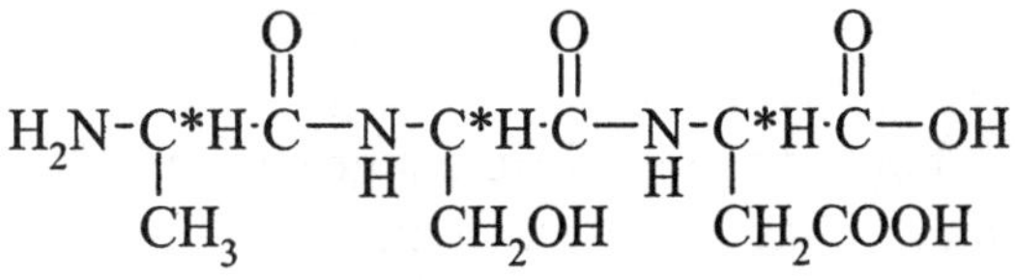

$$H_2N-\overset{}{\underset{|}{C}}^*H\cdot\overset{O}{\underset{||}{C}}-\underset{\underset{CH_3}{|}}{N}\cdot\overset{}{\underset{|}{C}}^*H\cdot\overset{O}{\underset{||}{C}}-\underset{\underset{CH_2OH}{|}}{N}\cdot\overset{}{\underset{CH_2COOH}{C}}^*H\cdot\overset{O}{\underset{||}{C}}-OH$$

A molecule with n tetrahedral stereocenters can form 2^n stereoisomers, thus this molecule can form eight stereoisomers.

17.33 A cyclic compound can exhibit chirality if a carbon in the ring has two different nonring substituents and the ring is not symmetrical with respect to that carbon.

(a) 3-methylcyclohexanone: C-3 has two different substituents and the ring is not symmetrical at C-3. Therefore, there are two optically active enantiomers:

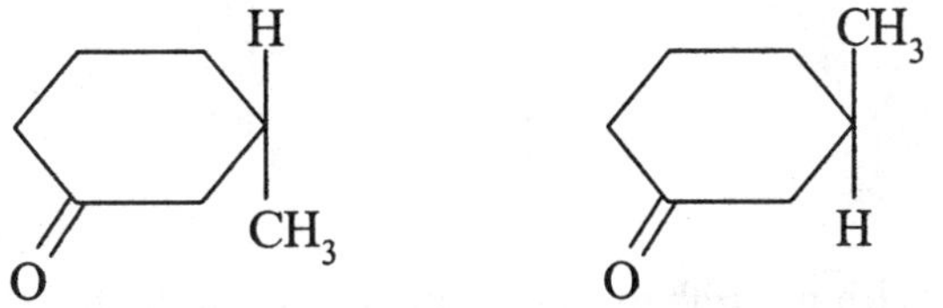

(b) 1,1dimethylcyclopentane: There is no carbon in the ring that has two different substituents, the carbons have either two hydrogens or in the case of C-1, two methyl groups. Thus there are no stereoisomers and this compound is not optically active.

(c) 2-methyl-1, 3-cyclohexanedione: This compound has a carbon, C-2, that has two different substituents; however the ring is symmetrical about this carbon, thus it is not optically active, and there are no stereoisomers.

(d) 1-hydroxy-1-methylcyclohexane: This compound has a carbon, C-1, that has two different substituents. However, the ring is symmetric about this carbon, thus it is not optically active, there are no stereoisomers.

(e) 1,2-dihydroxycyclohexane: This compound has two ring carbons each of which has two different substituents. Because the substituents on the two ring carbons have the same identities, there are three, not four stereoisomers. Inspection shows that there is one cis meso compound that is not optically active. The pair of trans compounds are enantiomers and are optically active.

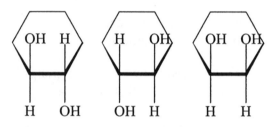

17.35 (a) Each carbon that is bonded to four different groups is a tetrahedral stereocenter. There are four such carbons in this molecule:

A molecule with n tetrahedral stereocenters can form 2^n stereoisomers, thus this molecule can form 16 stereoisomers.

(b) Each carbon that is bonded to four different groups is a tetrahedral stereocenter. There is one such carbon in this molecule:

A molecule with n tetrahedral stereocenters can form 2^n stereoisomers, thus this molecule can form two stereoisomers.

(c) Each carbon that is bonded to four different groups is a tetrahedral stereocenter. There are six such carbons in this molecule:

A molecule with n tetrahedral stereocenters can form 2^n stereoisomers, thus this molecule can form 64 stereoisomers.

17.37 When comparing two compounds, the first action should be to count the number of each type of atom to find a molecular formula. If they are not the same, then the two compounds must be different compounds that are not isomers. Inspection of the compounds in this question shows that the two compounds in (o) are different compounds ($C_7H_{13}Cl$ vs. C_8H_{16}) that are not isomers of each other.

The next step should be to determine if the two molecules are the same molecule. In this question, that can best be determined by superimposing the two molecules on top of each other. Care must be taken here to preserve the Fischer projections and it may simplify the comparison if both structures are written in a similar fashion. In (i) and (j), if either molecule is rotated 180° in the plane of the page, the other molecule will be formed, i.e. the two molecules become superimposable, in (b), (e), and (g) the two molecules are superimposable as written.

Further inspection differentiates between enantiomers, diastereomers, or constitutional isomers. In constitutional isomers, the connectivity of the atoms is different while enantiomers or diastereomers have the same connectivity, yet the two molecules cannot be superimposed. In (a), (d), and (m) the two compounds have the same molecular formula, but different connectivity and thus are constitutional isomers. Next, if two molecules are enantiomers, they must be mirror images of each other and have the same connectivity. In (c), it can be seen that the two molecules are mirror images if the first compound is placed above the second. In (h) and (l), the two molecules are mirror images if the mirror is placed between the two. Lastly, the compounds in (f), (k) and (n) have the same connectivity and yet are not mirror images of each other, thus they are diastereomers.

Thus:

(a) 2; (b) 1; (c) 3; (d) 2; (e) 1; (f) 4; (g) 1; (h) 3; (i) 1; (j) 1; (k) 4; (l) 3; (m) 2; (n) 4; (o) 5

17.39   4-chloro-2-pentene has one tetrahedral stereocenter at C-3. However, due to the double bond, there are two geometric stereoisomers (cis and trans) due to the presence of the double bond. Thus this compound can form four stereoisomers, a pair of cis enantiomers and a pair of trans enantiomers.

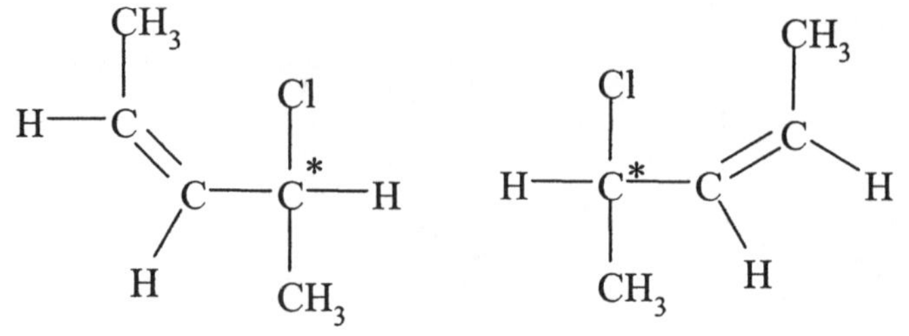

cis pair of isomers

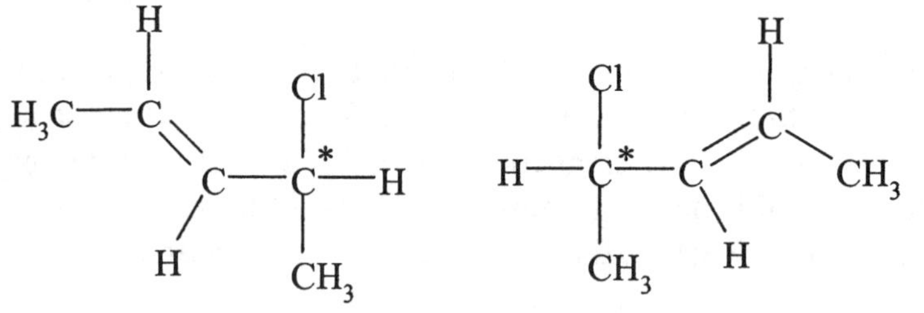

trans pair of isomers

17.41   Both reactants, HCl and 1-butene, are achiral and therefore the reaction cannot incorporate chiral recognition. Thus, there is an equal probability that the Cl will add to C2 from either side of the double bond to form equal amounts of both enantiomers of the product. Therefore, the product will be a racemic mixture of the two enantiomers of 2-chlorobutane that will not be optically active.

17.43   Both geometric isomers are achiral, and therefore chiral recognition cannot be responsible for the difference in the physiological effect of the two isomers. A similar mechanism to chiral recognition where a biological system preferentially recognizes one geometric isomer over the other is responsible for the different physiological response.

17.45   The square geometry would result in no enantiomers, but there would be three diastereomers:

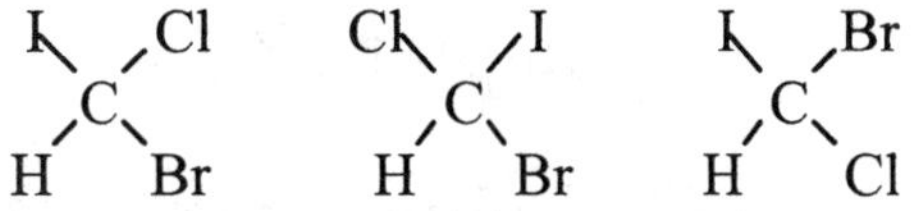

$$\text{I} \diagdown \diagup \text{Cl} \qquad \text{Cl} \diagdown \diagup \text{I} \qquad \text{I} \diagdown \diagup \text{Br}$$
$$\text{C} \qquad\qquad \text{C} \qquad\qquad \text{C}$$
$$\text{H} \diagup \diagdown \text{Br} \qquad \text{H} \diagup \diagdown \text{Br} \qquad \text{H} \diagup \diagdown \text{Cl}$$

The tetrahedral geometry would result in a pair of enantiomers

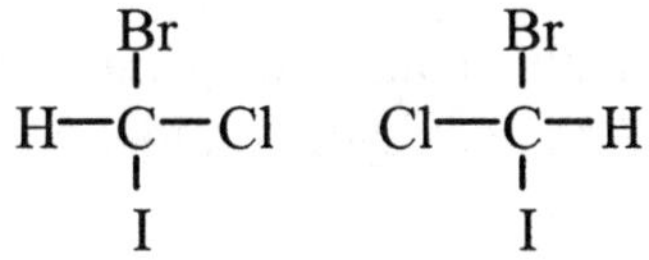

$$\begin{array}{ccc} & \text{Br} & & & \text{Br} & \\ & | & & & | & \\ \text{H}-&\text{C}&-\text{Cl} & \text{Cl}-&\text{C}&-\text{H} \\ & | & & & | & \\ & \text{I} & & & \text{I} & \end{array}$$

17.47 $\alpha = [\alpha]CL = 223° \times (0.100 \text{ g/mL}) \times 2.00 \text{ dm} = 44.6°$, is the expected specific rotation of an aqueous solution of (+)-penicillin V at a concentration of 0.100 g/mL, thus the concentration of the delivered product is not 0.100 g/mL (answer, part a). Its concentration is 0.0500 g/mL; $\alpha = [\alpha]CL = 223° \times (x \text{ g/mL}) \times 2.00 \text{ dm} = 22.3 \rightarrow x = 0.0500 \text{ g/mL}$.

17.49 There is no heteroatom substituent connected to its tetrahedral stereocenter, so the D/L system is not applicable to carvone. (-)– carvone is (R)–carvone.

17.51 (S)–methyldopa

17.53

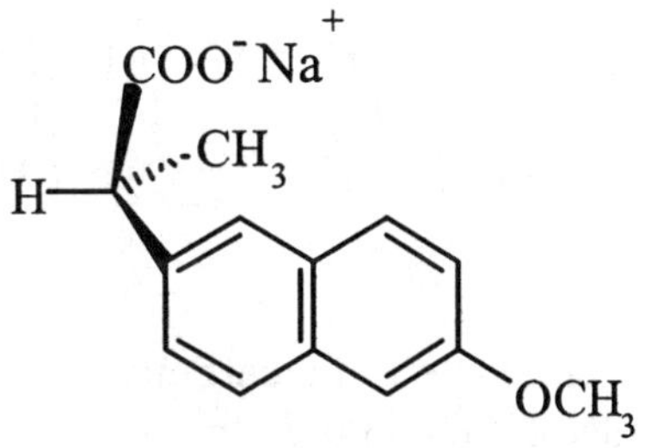

Chapter 18

Carbohydrates

18.1 (a) An aldose is a molecule with an aldehyde group and numerous hydroxy (–OH) groups. This is also called a polyhydroxyaldehyde.

(b) A hexose is an aldose or a ketose with six carbons. A ketose is a molecule with a ketone group and numerous hydroxy groups.

(c) A ketopentose is a ketose with five carbons. This may also be called a five-carbon polyhydroxyketone.

(d) An aldotetrose is an aldose with four carbons. This may also be called a four-carbon polyhydroxyaldehyde.

(e) All saccharides in the D family have the tetrahedral stereocenter farthest from the carbonyl carbon in the D configuration. In other words, the –OH group is on the right side of the carbon atom in the Fischer projection.

18.3 An aldopentose is a five-carbon molecule with hydroxy groups and an aldehyde. The structural formula that does not denote stereoisomerism is

$$\underset{CH_2OH-CH-CH-CH-CH}{\overset{OH\ \ OH\ \ OH\ \ O}{|\ \ \ \ |\ \ \ \ |\ \ \ \ ||}}$$

Therefore, the carbon at one end is part of a carbonyl, the carbon at the other end has two hydrogens and an –OH bonded to it, and the other three carbons are tetrahedral stereocenters. The presence of three tetrahedral stereocenters means that there exist $2^3 =$ 8 stereoisomers of this compound. One half of the stereoisomers are D and the other half are L, thus there are four L-aldopentoses.

18.5 A ketohexose is a six-carbon molecule with hydroxy groups and a ketone at C-2. Therefore, the carbons at both ends of the molecule have two hydrogens and an –OH bonded to it; one carbon is part of the carbonyl and the other three carbons are tetrahedral stereocenters. The presence of three tetrahedral stereocenters means that there exist $2^3 =$ 8 stereoisomers of this compound, four of which are D-ketohexoses. On the other hand, an aldohexose is a six-carbon molecule with hydroxy groups and an aldehyde. The carbon at one end is part of a carbonyl, the carbon at the other end has two hydrogens and an –OH bonded to it, and the other four carbons are tetrahedral stereocenters. The presence of four tetrahedral stereocenters means that there exist $2^4 = 16$ stereoisomers of this compound. One half of the stereoisomers are D and the other half are L, thus there are eight D-aldohexoses.

18.7   (a) Glucose is an aldehyde, while sorbose is a ketone, therefore these two compounds cannot be stereoisomers. They both have six carbons and thus are constitutional isomers.

(b) Fructose and sorbose are both ketohexoses and are therefore stereoisomers of each other; however, they are not mirror images of each other, therefore they are diastereomers.

(c) (+) Tagatose and (–) tagatose denote two molecules that only differ in their stereo arrangement and are mirror images of each other, therefore they are enantiomers.

(d) D-sorbose and L-sorbose denote two molecules that only differ in their stereo arrangement and are mirror images of each other, therefore they are enantiomers of each other.

(e) 2-deoxy-D-ribose has one less oxygen than D-ribose, and therefore these two molecules are different compounds that are not isomers.

18.9   Haworth projections are used to represent cyclic structures of monosaccharides.

18.11   A pyranose ring is a six-membered cyclic hemiacetal ring.

18.13   The difference between the α and β anomers in D-glucose is the relative positions of the –CH$_2$OH on C5 and the –OH on C1. In the α-anomer, these two groups are situated trans to each other, while in the β-anomer, they have a cis relationship.

18.15   The name allose denotes that the ring has five carbons and an oxygen and that the –OH groups on C2, C3, and C4 are all on the same side of the ring, opposite to the –CH$_2$OH on C5. D-allose denotes that all of the –CH$_2$OH on C5 is above the ring in the Haworth projection. Finally, α- D-allose denotes that the –CH$_2$OH group on C5 is trans to the –OH on C1. Thus the structure is:

18.17   The name psicose denotes that the ring has four carbons and an oxygen and that the –OH groups on C3 and C4 are on the opposite side of the ring as the –CH$_2$OH on C5. D-psicose denotes that the –CH$_2$OH on C5 is above the ring in the Haworth projection.

Finally, β- D-psicose denotes that the -CH$_2$OH group on C5 is cis to the –OH on C2. Thus, the structure is:

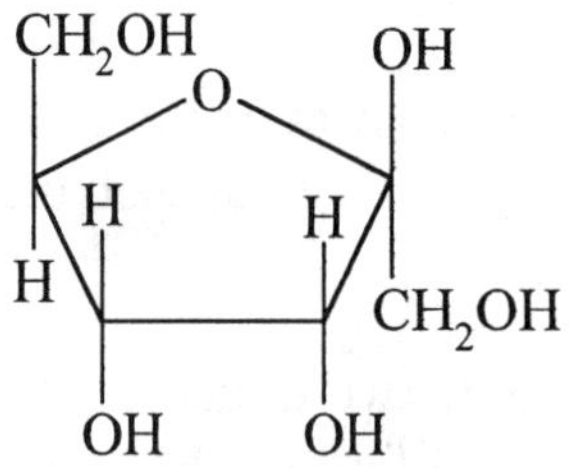

18.19   In solution, some of the cyclic acetal structure of β-D-glucose will open to form the linear aldehyde. The aldehyde structure then is present to give a positive Benedict's test. A positive test with Benedict's reagent is indicated by the disappearance of the blue color of Benedict's reagent and the formation of a red precipitate of Cu$_2$O.

18.21   Examination of Figures 18.1 and 18.2 shows that

(a) D-glyceraldehyde has only one carbon that is a tetrahedral stereocenter.

(b) D-sorbose has three carbons that are tetrahedral stereocenters.

(c) L-mannose has four carbons that are tetrahedral stereocenters.

(d) 2-deoxy- D-ribose. In a "deoxy" compound, one of the –OH groups is replaced with a –H, thus that carbon is no longer a stereocenter. Ribose has three carbons that are tetrahedral stereocenters, therefore 2-deoxy- D-ribose has two carbons that are tetrahedral stereocenter.

18.23   The hemiacetal in glucose can undergo dehydration in the presence of an alcohol to form a glycoside as:

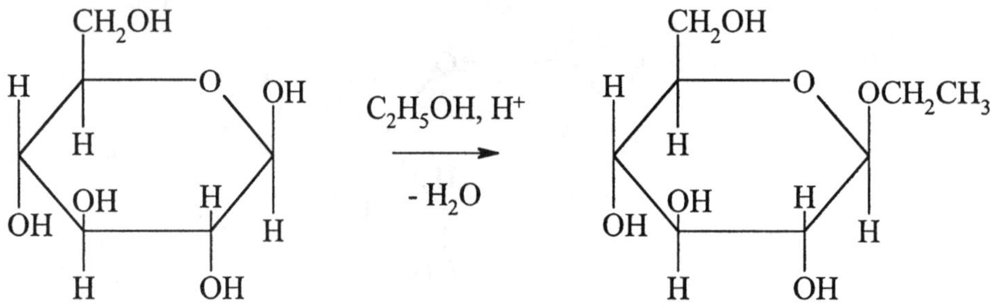

18.25   Reaction with Benedict's reagent results in an oxidation. In monosaccharides, the aldehyde undergoes oxidation to form a carboxylic acid. Therefore, D-galactose will undergo oxidation to form:

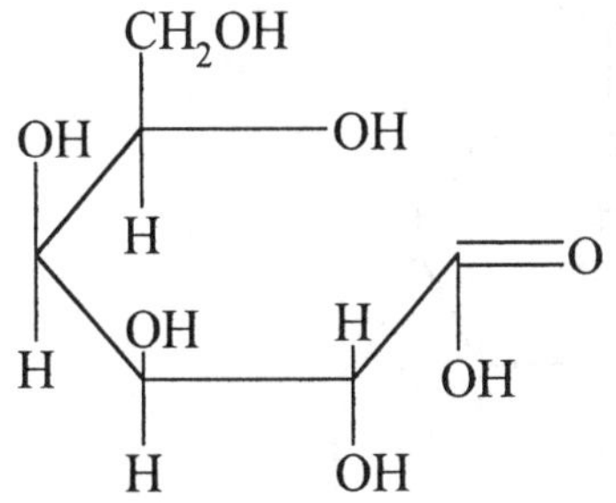

$$
\begin{array}{c}
CH_2OH \\
\end{array}
$$

OH ⎯⎯⎯ OH

H

OH   H   ═O

H   OH

H   OH

18.27 A reducing sugar is one that will form an open chain structure that will react with an oxidizing agent such as Benedict's reagent. To do this it must have a hemiacetal linkage in the ring. Of the compounds listed in this problem, all except B can undergo this reaction and are thus reducing agents and undergo mutarotation.

18.29 The $\alpha(1 \rightarrow 4)$ glycosidic linkage connects two aldohexose rings by creating a bond between the C1 carbon in the α position of one ring to the C4 position of the other.

18.31 (a) Maltose is two glucose rings bonded together by a glycosidic linkage. Digestion breaks this glycosidic linkage to leave two glucose rings.

(b) Lactose is a glucose ring bonded to a galactose ring by a glycosidic linkage. Digestion breaks this glycosidic linkage to leave a glucose and a galactose ring.

(c) Sucrose is a glucose ring bonded to a fructose ring by a glycosidic linkage. Digestion breaks this glycosidic linkage to leave a glucose and a fructose ring.

(d) Humans cannot digest cellobiose.

18.33   Figure 18.8 in the text shows the structure of sucrose and is shown below:

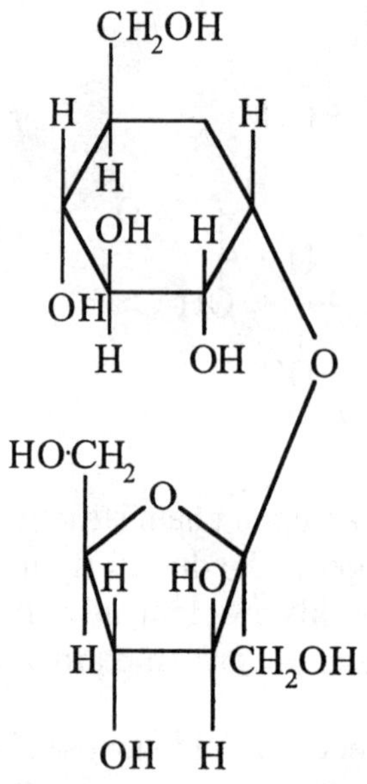

Sucrose is not a reducing sugar because it does not contain a hemiacetal group that can open to the aldehyde structure and undergo oxidation.

18.35   Two glucose rings linked by a β(1 → 6) linkage are:

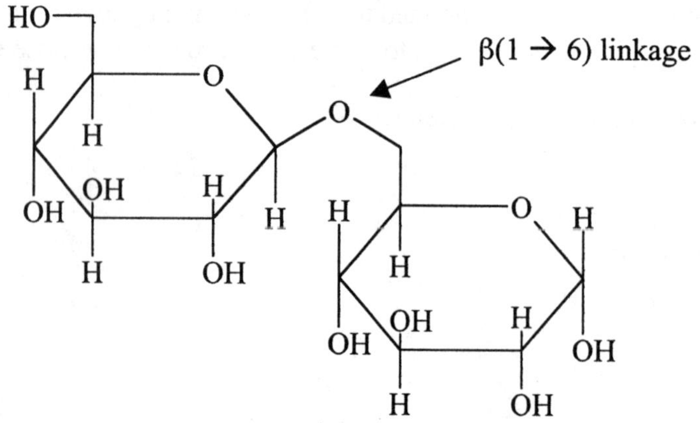

Gentiobiose is a reducing sugar because it contains a hemiacetal group in the ring that can open to the aldehyde structure and undergo oxidation.

18.37   Sucrose is not a reducing sugar and thus will not give a positive Benedict's test. Lactose is a reducing sugar and will give a positive Benedict's test. As the unknown does not give a positive Benedict's test, it must be sucrose.

18.39   Both amylopectin and amylose are polysaccharides that have many D-glucose rings linked together by $\alpha(1 \rightarrow 4)$ glycosidic linkages. However, amylose is a linear molecule but amylopectin is branched repeatedly by $\alpha(1 \rightarrow 6)$ glycosidic linkages. Starch is a mixture of amylose and amylopectin that is found in plants as a method to store D-glucose.

18.41   Humans can digest amylose, amylopectin, and glycogen to break the glycosidic linkages and yield D-glucose; humans cannot digest cellulose. Cows digest amylose, amylopectin, and cellulose to break the glycosidic linkages and yield D-glucose. Because they are herbivores and glycogens are only present in animals, grazing animals do not ingest glycogen.

18.43   N-acetyl- D-glucosamine is a modified glucose unit. Chitin is this unit bonded by $\beta(1 \rightarrow 4)$ glycosidic linkages, thus the structure is:

18.45   Plants provide the immediate source of organic compounds for animals. Photosynthesis is the ultimate source of organic compounds for animals because plants utilize photosynthesis to create organic compounds.

18.47   A carbohydrate is a compound that is (or is derived from) an aldose or ketose. An aldose is a molecule with an aldehyde group and numerous hydroxy (–OH) groups. This is also called a polyhydroxyaldehyde. A ketose is a molecule with a ketone group and numerous hydroxy (–OH) groups. This is also called a polyhydroxyketone.

18.49   Only monosaccharides and disaccharides are called sugars.

18.51   In the Fischer projection, the placement of the –OH on the tetrahedral stereocenter farthest from the carbonyl (C5) is on the right side in the D configuration. In the Haworth projection the –CH$_2$OH on C5 extends upward from the plane of the ring.

18.53   (a) Sucrose is two glucose rings bonded together and is therefore a disaccharide.

(b) Galactose is a single aldohexose and is therefore a monosaccharide.

(c) Lactose is a glucose and a galactose ring linked together and is therefore a disaccharide.

(d) Amylose is a compound that has many D-glucose rings linked together by α(1 → 4) glycosidic linkages. Therefore, it is a polysaccharide.

(e) Amylopectin is a compound that has many D-glucose rings linked together by α(1 → 4) glycosidic linkages and also branches repeatedly by α(1 → 6) glycosidic linkages. Therefore, it is a polysaccharide.

(f) Fructose is a single ketohexose and is therefore a monosaccharide.

(g) Cellulose is a linear molecule that has many D-glucose units connected together by β(1 → 4) glycosidic linkages. Therefore, it is a polysaccharide.

18.55   (a) α- D-glucose and β- D-glucose denote two molecules that only differ in the way that the –OH on C1 is bonded. As all of the atoms are bonded the same way in this compound, they are stereoisomers of each other. However, they are not mirror images, and therefore are diastereomers.

(b) Galactose and fructose are both hexoses, but fructose is a ketone while galactose is an aldehyde, therefore they cannot be stereoisomers. They both have six carbons and thus are constitutional isomers

(c) (+) Galactose and (–) galactose denote two molecules that only differ in their stereo arrangement and are mirror images of each other, therefore they are enantiomers.

(d) Maltose and cellobiose are disaccharides that consist of two glucose molecules and therefore have the same molecular formula. However, different linkages in the two compounds bond the two rings. Therefore, the two compounds are not stereoisomers, but are constitutional isomers.

(e) Amylose is a polysaccharide while glucose is a monosaccharide. Therefore, amylose has many more carbon atoms than glucose and therefore these two compounds are different compounds that are not isomers.

(f) Maltose is a disaccharide while glucose is a monosaccharide. Therefore, Maltose has twice as many carbon atoms than glucose and therefore these two compounds are different compounds that are not isomers.

(g) Cellulose is a linear polysaccharide that has many D-glucose rings linked together by β(1 → 4) glycosidic linkages. Amylose is also a linear polysaccharide that has many D-glucose rings linked together by α(1 → 4) glycosidic linkages. Therefore, they have the same molecular formula, but different connectivity and are therefore constitutional isomers.

18.57   Storage polysaccharides are compounds that have many units (usually D-glucose) linked together by glycosidic linkages. An outside body can then break these linkages to yield the units such as D-glucose that can then be used to generate energy or synthesize other compounds. Structural polysaccharides, however, are used to construct cell walls in plants. These walls then impart stability and macroscopic shape to the plant.

18.59   Grazing animals like cows have microorganisms in their digestive tract that possess cellulase, the required enzyme to break the β(1 → 4) glycosidic linkages in cellulose. When the microorganism breaks these linkages, D-glucose remains, which the animals use for energy. Grazing animals themselves have amylase and maltase, the enzymes needed to break the α(1 → 4) glycosidic linkage of starch during digestion. This decomposition also results in glucose that the animal can use for energy.

18.61   Hydrolysis breaks the glycosidic linkages between monosaccharides, thus if one mole of carbohydrate yields four moles of monosaccharide, the carbohydrate must be made of four monosaccharides. In other words, it is a tetrasaccharide.

18.63   The structure of isomaltose is:

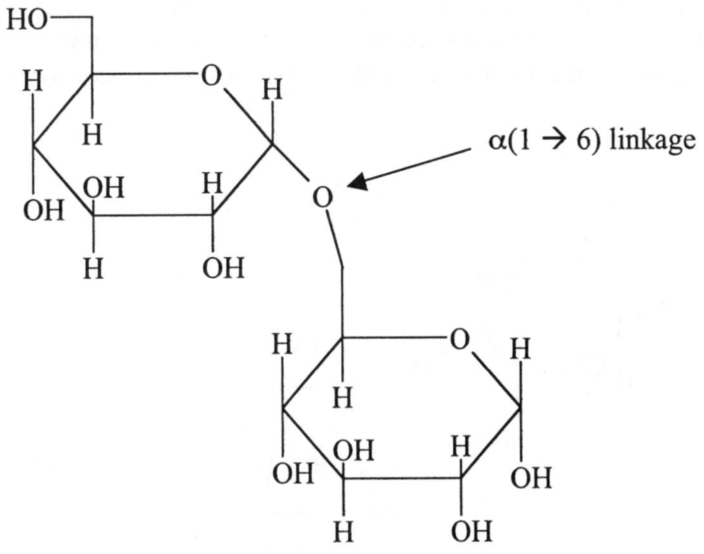

The presence of a hemiacetal group on the bottom ring in this structure means that isomaltose is a reducing sugar and undergoes mutarotation.

18.65 $6 CO_2 + 6 H_2O + sunlight + chlorophyll → (CH_2O)_6 + 6 O_2$

18.67 This is the cycle that oxygen undergoes as it is used and produced by plants. Plants produce oxygen as a result of photosynthesis. Plants and animals also use oxygen in metabolism to generate energy.

18.69 The glucose residue $C_6H_{10}O_5$ has a molecule mass of 162 amu. The number of residues in a polysaccharide is the molecular mass of the polysaccharide divided by 162. Therefore, there are 1000 glucose units on either amylose or cellulose that has a molecular mass of 162,000.

18.71 This stereoselective linkage must be formed by chiral recognition whereby an enzyme that allows only the two correct stereoisomers to come together in a specified manner to form the correct glycosidic linkage.

18.73 Lactose intolerance is a result of the absence of the proper enzyme, lactase, needed to digest lactose. Symptoms of lactose intolerance include abdominal distention and cramping, nausea, pain and diarrhea. These symptoms are temporary, however, and will subside in the absence of lactose. Galactosemia has long-term adverse effects if not detected early in infancy. Long-term adverse effects include mental retardation, impaired liver function, cataracts, and death.

18.75 Saccharin is 300 times sweeter than sucrose, therefore, 8.0/300 = 0.027 g of saccharin (27 mg) is equivalent to 8.0 g of sucrose in sweetness. Thus the manufacturers instructions suggest that you need more (35 mg) than the 27 mg that Box 18.1 indicates. The manufacturers directions may be a method to insure more saccharin is sold or it may be a method to insure that the user is satisfied with the sweetness of saccharin when it replaces sugar.

18.77

18.79 People with AB blood type do not produce antibodies (neither anti-A or anti-B) because they have A and B antigens on their red blood cell surfaces. Since the only types of antigens that are present in any blood type are A and B, no blood type produces an antibody-antigen reaction in a person with type AB blood.

18.81 The daily consumption of sucrose (40 g) contains 160 kcal:

$$5 \text{ cups of coffee} \times \frac{2 \text{ teaspoons of sugar}}{\text{cup of coffee}} \times \frac{4.0 \text{ g sugar}}{\text{teaspoon of sugar}} \times \frac{4.0 \text{ kcal}}{\text{g sugar}} = 160 \text{ kcal}$$

As aspartame is 180 times sweeter than sugar (Box 18.2), the 40 g of sugar needed to sweeten the five cups of coffee can be replaced by 40/180 = 0.22 g aspartame, which contains 0.88 kcal

$$0.22 \text{ g aspartame} \times \frac{4.0 \text{ kcal}}{\text{g aspartame}} = 0.88 \text{ kcal}$$

Thus, replacing sugar with aspartame will save 160 kcal – 0.88 kcal = 159 kcal.

18.83   Acid or sucrase will allow the sucrose molecule to undergo hydrolysis, which results in a mixture of fructose and glucose, both of which have different specific rotation values from sucrose.

18.85

18.87   The α-linkages between glucose residues in a starch molecule result in the molecule forming a helical conformation, which contains intramolecular hydrogen bonds between different parts of the molecule. This conformation does not allow, however, many intermolecular hydrogen bonds, which allows the molecule to form many hydrogen bonds with water, when put in contact with water, which provides a mechanism for dissolution of starch in water. Cellulose, however, contains mostly β-linkages between glucose residues, which results in an extended chain conformation, and produces extensive hydrogen bonds among different chains and very little hydrogen bonding between cellulose and water. This results in very little interaction between the water and cellulose molecules and poor solubility of cellulose in water.

18.89   Both D-glucose and D-fructose have the same molecular formula, $C_6H_{12}O_6$, and thus the balanced equation for their combusion is identical:

$$C_6H_{12}O_6 + 6 O_2 \rightarrow 6 CO_2 + 6 H_2O$$

18.91   One reaction is the exact reverse of the other. The path by which the two reactions occur are completely different, but the chemical reactions are exactly the reverse of each other. Combustion requires the presence of oxygen and an ignition source, such as a flame or spark, while photosynthesis requires light and chlorophyll.

Chapter 19

Lipids

19.1 The fatty acids found in animals and plants have three particular characteristics; they have an even number of carbons; are unbranched; and when double bonds are present, they have a cis configuration. In this question only compound 1 fulfills all of these requirements. Compound 2 is branched, while compound 3 has an odd number of carbons.

19.3 Saturated fatty acids have no carbon-carbon double bonds. An unsaturated fatty acid will contain one or more carbon-carbon double bond in the cis configuration. Unsaturated fatty acids are further broken up into monosaturated and polyunsaturated fatty acids. Monosaturated fatty acids contain only one carbon-carbon double bond, while polyunsaturated fatty acids contain more than one carbon-carbon double bond.

19.5 Compound 2 has a higher melting point because molecules of 2 can pack more tightly together due to the flexible linear chain of single bonds. The double bond in compound 1 causes the chain to kink and thus the chains cannot pack as tightly as the saturated chain. This closer packing of compound 2 results in stronger intermolecular interactions that result in a higher melting point.

19.7 A triacylglycerol is the triester of glycerol. Lauric acid is a saturated acid with 12 carbons, myristic acid is a saturated acid with 14 carbons, and oleic acid is a monosaturated acid with 18 carbons. The triacylglycerol is created by removing the –OH from the acid and an –H from a hydroxyl group on glycerol, then creating a bond between the carbonyl of the acid to the –O– of glycerol. Thus this compound is:

$$
\begin{array}{l}
CH_2\!-\!O\!-\!\overset{\displaystyle O}{\overset{\|}{C}}\!-\!(CH_2)_{10}CH_3 \\[2pt]
CH\!-\!O\!-\!\overset{\displaystyle O}{\overset{\|}{C}}\!-\!(CH_2)_{12}CH_3 \\[2pt]
CH_2\!-\!O\!-\!\overset{\displaystyle O}{\overset{\|}{C}}\!-\!(CH_2)_7\!-\!CH\!=\!CH(CH_2)_7CH_3
\end{array}
$$

19.9 Melting point decreases as the amount of double bonds in the triacylglycerol increases. Compound a has no double bonds, compound b has three double bonds and compound c has one double bond. Thus, in increasing order of melting point, compound b < compound c < compound a.

19.11 (a) Saponification is the reaction of the ester group with water in the presence of a base.

The result is that the ester cleaves to form the alcohol and the carboxylate salt. In triacylglycerols, this reaction occurs at each ester to form:

$$
\begin{array}{ccc}
\underset{\displaystyle CH_2\text{-}O\text{-}\overset{\displaystyle O}{\overset{\|}{C}}\text{-}(CH_2)_{12}CH_3}{}\\
\underset{\displaystyle CH\text{-}O\text{-}\overset{\displaystyle O}{\overset{\|}{C}}\text{-}(CH_2)_{14}CH_3}{}\\
\underset{\displaystyle CH_2\text{-}O\cdot\overset{\displaystyle O}{\overset{\|}{C}}\text{-}(CH_2)_7CH\!=\!CH(CH_2)_7CH_3}{}
\end{array}
\quad\xrightarrow[\text{NaOH}]{\text{H}_2\text{O}}\quad
\begin{array}{c}
CH_2\text{—OH}\\
CH\text{—OH}\\
CH_2\text{—OH}
\end{array}
\;+\;
\begin{array}{c}
NaO\text{-}\overset{O}{\overset{\|}{C}}\text{—}(CH_2)_{12}CH_3\\
NaO\text{-}\overset{O}{\overset{\|}{C}}\text{—}(CH_2)_{14}CH_3\\
NaO\text{-}\overset{O}{\overset{\|}{C}}\text{-}(CH_2)_7\text{—}CH\!=\!CH(CH_2)_7CH_3
\end{array}
$$

(b) Catalytic hydrogenation in the presence of Pt results in the addition of H_2 across the carbon-carbon double bond. The only place that this can occur in this compound is in the bottom fatty acid as:

$$
\begin{array}{c}
CH_2\text{—}O\text{-}\overset{O}{\overset{\|}{C}}\text{-}(CH_2)_{12}CH_3\\
CH\text{—}O\text{-}\overset{O}{\overset{\|}{C}}\text{-}(CH_2)_{14}CH_3\\
CH_2\text{—}O\text{-}\overset{O}{\overset{\|}{C}}\text{-}(CH_2)_7\text{—}CH\!=\!CH(CH_2)_7CH_3
\end{array}
\quad\xrightarrow[\text{Pt}]{\text{H}_2}\quad
\begin{array}{c}
CH_2\text{—}O\text{—}\overset{O}{\overset{\|}{C}}\text{—}(CH_2)_{12}CH_3\\
CH\text{—}O\text{—}\overset{O}{\overset{\|}{C}}\text{—}(CH_2)_{14}CH_3\\
CH_2\text{—}O\text{—}\overset{O}{\overset{\|}{C}}\text{-}(CH_2)_{16}CH_3
\end{array}
$$

19.13   Bacterial hydrolysis of this compound breaks the ester linkages to form fatty acids and glycerol. One of the fatty acids that result is butanoic acid, $C_4H_8O_2$. Because of its low molar mass, butanoic acid is volatile and produces a foul odor.

$$
\begin{array}{c}
CH_2\text{—}O\text{—}\overset{O}{\overset{\|}{C}}\text{-}(CH_2)_2CH_3\\
CH\text{—}O\text{—}\overset{O}{\overset{\|}{C}}\text{-}(CH_2)_{14}CH_3\\
CH_2\text{—}O\text{—}\overset{O}{\overset{\|}{C}}\text{-}(CH_2)_7\text{—}CH\!=\!CH(CH_2)_5CH_3
\end{array}
\quad\xrightarrow{\text{H}_2\text{O}}\quad
\begin{array}{c}
CH_2\text{—OH}\\
CH\text{—OH}\\
CH_2\text{—OH}
\end{array}
\;+\;
\begin{array}{c}
HO\text{—}\overset{O}{\overset{\|}{C}}\text{—}(CH_2)_2CH_3\\
HO\text{—}\overset{O}{\overset{\|}{C}}\text{—}(CH_2)_{14}CH_3\\
HO\text{—}\overset{O}{\overset{\|}{C}}\text{-}(CH_2)_7\text{—}CH\!=\!CH(CH_2)_5CH_3
\end{array}
$$

19.15   By hydrogenation of the carbon-carbon double bonds in corn oil in the presence of a metal catalyst, the fatty acids become more saturated, and thus linear. These linear chains pack more readily, which increases the melting point of the compound.

19.17   The waxes found in nature tend to be an ester that is the combination of a carboxylic acid and an alcohol. The acid and alcohol normally contain between 14 and 36 carbons each, are unbranched, and have an even number of carbons. Therefore, the structure of the wax should have an even number of carbons, between 14 and 36, on both sides of the oxygen that is next to the carbonyl and be linear. Compound 1 has only four carbons to the left of the carbonyl, while compound 3 has an odd number of carbons on both sides of the oxygen, thus compound 2 is the correct compound.

19.19   Both triacylglycerols and glycerophospholipids are created by forming ester linkages to the three –OH groups of glycerol. In triacylglycerol, all three esters are bonded to fatty acids. However in glycerophospholipids, two of the ester links are bonded to fatty acids, but the third is linked to a phosphodiester group.

19.21   If the hydrolysis yields myristic acid and oleic acid, these two fatty acids must be bonded to two of the ester groups of glycerol. The presence of phosphoric acid and serine in the products tells us that the phosphodiester is bonded to serine (from Table 19.3). Thus, the structure is:

$$CH_2-O-\overset{\overset{\displaystyle O}{\|}}{C}-(CH_2)_{12}CH_3$$
$$CH-O-\overset{\overset{\displaystyle O}{\|}}{C}-(CH_2)_7-CH=CH(CH_2)_7CH_3$$
$$CH_2-O-\overset{\overset{\displaystyle O}{\|}}{\underset{\underset{\displaystyle O^-}{|}}{P}}-O-CH_2\overset{\overset{\displaystyle +}{}}{C}H-NH_3$$
$$COO^-$$

19.23   The hydrolysis breaks the top two ester groups to form –OH on glycerol and two fatty acids. The phosphodiester decomposes to phosphoric acid and the alcohol $HOCH_2CH_2NH_3^+$:

$$CH_2-OH$$
$$CH-OH$$
$$CH_2-OH$$

$$HO-\overset{\overset{\displaystyle O}{\|}}{C}-(CH_2)_{14}CH_3$$
$$HO-\overset{\overset{\displaystyle O}{\|}}{C}-(CH_2)_{16}CH_3$$
$$HO-\overset{\overset{\displaystyle O}{\|}}{\underset{\underset{\displaystyle O^-}{|}}{P}}-OH \qquad HO-CH_2CH_2NH_3^+$$

19.25   Linoleic acid is a polyunsaturated acid with 18 carbons, and choline is shown in Table 19.3. Linoleic acid will combine at the amine in sphingosine to form an amide group, while the phosphoric acid and choline combine at the –OH of sphingosine to form:

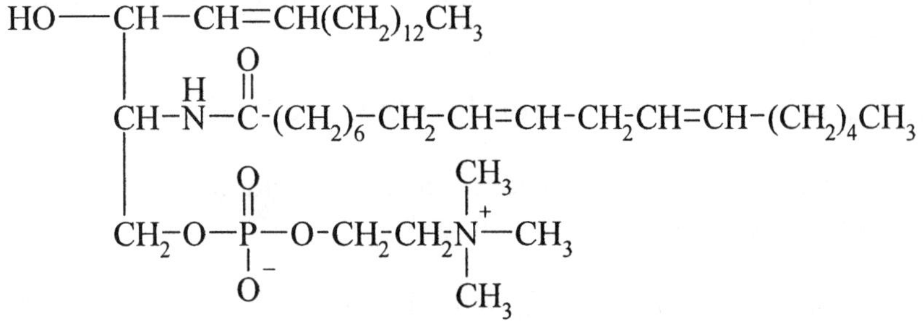

$$\text{HO}-\text{CH}-\text{CH}=\text{CH(CH}_2)_{12}\text{CH}_3$$

$$\underset{\overset{|}{\text{CH}}}{}-\overset{\text{H}}{\underset{}{\text{N}}}-\overset{\overset{\text{O}}{\|}}{\text{C}}\text{-(CH}_2)_{\overline{6}}\text{CH}_{\overline{2}}\text{CH}=\text{CH-CH}_2\text{CH}=\text{CH-(CH}_2)_4\text{CH}_3$$

$$\text{CH}_{\overline{2}}\text{O}-\overset{\overset{\text{O}}{\|}}{\underset{\underset{\text{O}^-}{|}}{\text{P}}}-\text{O-CH}_{\overline{2}}\text{CH}_{\overline{2}}\overset{\overset{\text{CH}_3}{|}}{\underset{\underset{\text{CH}_3}{|}}{\text{N}^+}}-\text{CH}_3$$

19.27 An acid catalyzed hydrolysis of this lipid breaks the amide and ester linkages to form sphingosine, oleic acid, phosphoric acid, and choline:

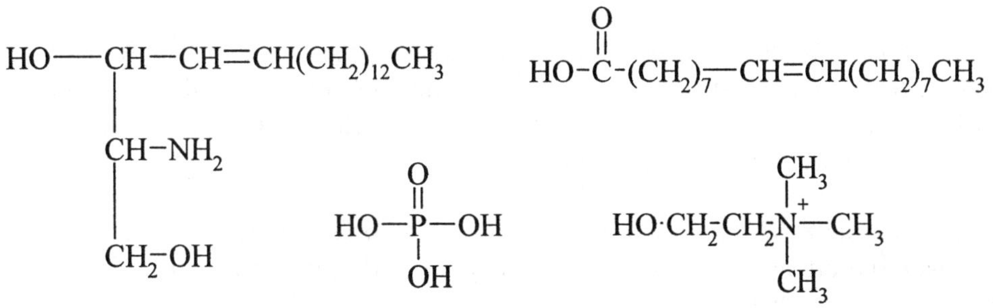

$$\text{HO}-\text{CH}-\text{CH}=\text{CH(CH}_2)_{12}\text{CH}_3 \qquad \text{HO-}\overset{\overset{\text{O}}{\|}}{\text{C}}\text{-(CH}_2)_{\overline{7}}-\text{CH}=\text{CH(CH}_2)_7\text{CH}_3$$

$$\overset{|}{\underset{\overset{|}{\text{CH}_{\overline{2}}\text{OH}}}{\text{CH}-\text{NH}_2}}$$

$$\text{HO}-\overset{\overset{\text{O}}{\|}}{\underset{\underset{\text{OH}}{|}}{\text{P}}}-\text{OH} \qquad \text{HO·CH}_{\overline{2}}\text{CH}_{\overline{2}}\overset{\overset{\text{CH}_3}{|}}{\underset{\underset{\text{CH}_3}{|}}{\text{N}^+}}-\text{CH}_3$$

19.29 Steroids do not contain functional groups such as an ester or amide group that can be cleaved during hydrolysis to form smaller molecules.

19.31 A tetrahedral stereocenter is one that is bonded to four different groups. In testosterone there are six such carbons, denoted by asterisks below. As there are six tetrahedral stereocenters, there are $2^6 = 64$ stereoisomers of testosterone.

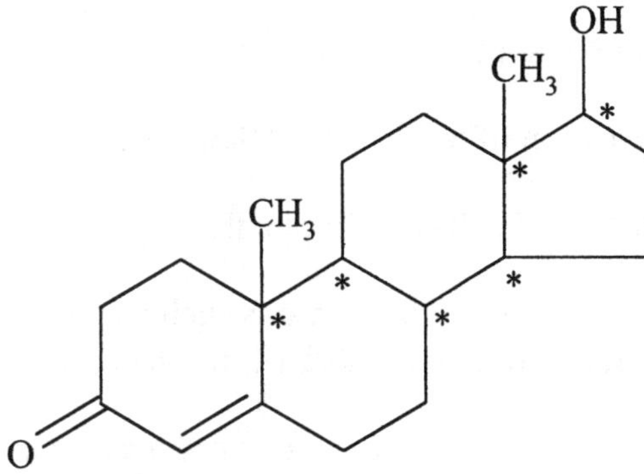

19.33 The sex hormones regulate the development of the sex organs, production of sperm and ova. They also regulate the development of secondary sex characteristics: lack of facial hair, increased breast size, and high voice in women; facial hair, increased musculature, and deep voice in men.

19.35   Cholesterol is the precursor to steroid hormones and bile salts.

19.37   Bile salts are compounds that replace a carboxylic acid group that is bonded to the ring structure in a steroid with a carboxylate salt.

19.39   Hydrolysis can occur at the amide group. However, the portion of the molecule that is removed is small compared to the overall structure of leukotriene D_4. Thus this change does not substantially alter the size of the molecule.

19.41   In leukotriene, the 20-carbon chain of arachidonic acid and the carboxyl group remain intact. In prostaglandin, a cyclopentane ring is formed between the C8 and C12 of the 20-carbon chain.

19.43   Hormones are synthesized in and secreted by endocrine glands and then transported in the bloodstream to target tissues where they regulate the functions of cells. Eicosanoids are called local hormones because they are not transported in the blood stream, but act in the same tissue that they were synthesized.

19.45   Vitamins are organic compounds that are required in trace amounts for normal metabolism, but are not synthesized by the organism that requires them. They, therefore, must be included in diet.

19.47   Fat-soluble vitamins dissolve in fat, but not water; water-soluble vitamins dissolve in water, but not fat.

19.49   A complete description of these functions can be found in Table 19.4.

Vitamin A: Plays a key role in vision by aiding proper function of mucous membranes.

Vitamin D: Regulates calcium and phosphate use and deposition in bone and cartilage.

Vitamin E: Acts as an antioxidant to protect cell membrane lipids from cleavage by O_2.

Vitamin K: Regulates formation of prothrombin that is needed for blood clotting.

19.51   Membranes form the walls of all cells and organelles.

19.53   Lipids with hydrophilic and hydrophobic portions such as glycerophospholipids, sphingolipids, or cholesterol form the membrane into the lipid bilayer.

19.55   Unsaturated fatty acids have cis double bonds that produce kinks in the chains. This results in a lipid bilayer that is not tightly packed, that has weaker secondary forces, and more flexibility.

19.57   The ions and D-glucose are very hydrophilic (they like water). Therefore, the hydrophobic regions of the lipid bilayer will repel them.

19.59   Active transport is the diffusion of a compound from an area of low concentration to an area of high concentration. This transport against a concentration gradient can only happen if energy is added to the system.

19.61   Because lactose is being transported from a region of low concentration to one of high concentration, this must be active transport.

19.63   Lipids are not very polar; therefore, the most polar molecule, CH_3OH will be the least effective at dissolving lipids.

19.65   (a) This compound does not have a carbon-carbon double bond and therefore it does not have a ω number.

(b) The first double bond is seven carbons from the methyl endgroup of this acid, thus it is ω-7.

(c) The first double bond is six carbons from the methyl endgroup of this acid, thus it is ω-6.

19.67   Waxes form a protective coating on plants and animals. They protect against parasites, excessive water loss, mechanical damage, and waterproof fowl.

19.69   Digestion does not result in the cleavage of all the hydrolyzable groups. A mixture of products is usually obtained. For instance triacylglycerols usually yield monoacylglycerols and fatty acids. Basic hydrolysis yields fatty acid salts instead of fatty acids.

19.71   A lipid is hydrolyzable if it contains an ester or amide group that can be cleaved in the presence of water. Of the compounds listed here, only (a) sphingolipids, (c) glycerophospholipids, (f) triacylglycerols, and (g) waxes contain these groups and can undergo hydrolysis.

19.73   A lipid can undergo saponification if it contains an ester group that can be cleaved in the presence of a base and water. Of the compounds listed here, only (a) sphingolipids, (c) glycerophospholipids, (f) triacylglycerols, and (g) waxes contain esters and can undergo saponification.

19.75   The three constitutional isomers will differ in the position at which the fatty acids are bonded to the glycerol, thus they are:

$$CH_2-O-\overset{\overset{\displaystyle O}{\|}}{C}-(CH_2)_{12}CH_3$$
$$|$$
$$CH-O-\overset{\overset{\displaystyle O}{\|}}{C}-(CH_2)_{14}CH_3$$
$$|$$
$$CH_2-O-\overset{\overset{\displaystyle O}{\|}}{C}-(CH_2)_{16}CH_3$$

$$CH_2-O-\overset{\overset{\displaystyle O}{\|}}{C}-(CH_2)_{12}CH_3$$
$$|$$
$$CH-O-\overset{\overset{\displaystyle O}{\|}}{C}-(CH_2)_{16}CH_3$$
$$|$$
$$CH_2-O-\overset{\overset{\displaystyle O}{\|}}{C}-(CH_2)_{14}CH_3$$

$$CH_2-O-\overset{\overset{\displaystyle O}{\|}}{C}-(CH_2)_{14}CH_3$$
$$|$$
$$CH-O-\overset{\overset{\displaystyle O}{\|}}{C}-(CH_2)_{12}CH_3$$
$$|$$
$$CH_2-O-\overset{\overset{\displaystyle O}{\|}}{C}-(CH_2)_{16}CH_3$$

19.77   The carbon-carbon double bond of oleic acid in triacylglycerol B yields a carboxylic acid when it undergoes air oxidation. This carboxylic acid has a bad odor and has a low enough molar mass to be volatile. Triacylglycerol A, however, does not have a carbon-carbon double bond and therefore will not undergo air oxidation.

19.79   A tetrahedral stereocenter is one that is bonded to four different groups. In progesterone there are six such carbons, denoted by asterisks below. As there are six tetrahedral stereocenters, there are $2^6 = 64$ stereoisomers of progesterone.

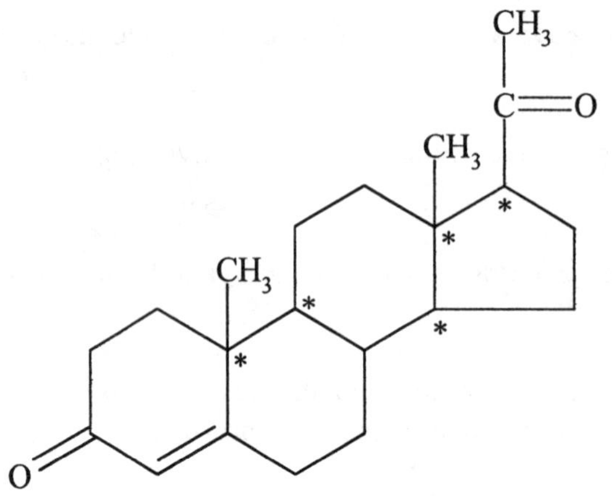

19.81   Triacylglycerols are completely hydrophobic; they have no parts that like water or are hydrophilic.

19.83   The fact that the protons are diffusing from an area of low concentration to high concentration means that this is active transport.

19.85   Coconut oil has fatty acid with very few double bonds. This usually results in a high melting point because the chains can pack well. However, the fatty acids in coconut oil have low molecular mass, usually less than 14 carbons per fatty acid. Thus, coconut oil is liquid at room temperature even though it is mostly saturated because it has a low melting point as a result of the low molecular mass of the fatty acids.

19.87   The ester that is formed is created by the reaction of the acid group of linoleic acid and the –OH group of cholesterol to form:

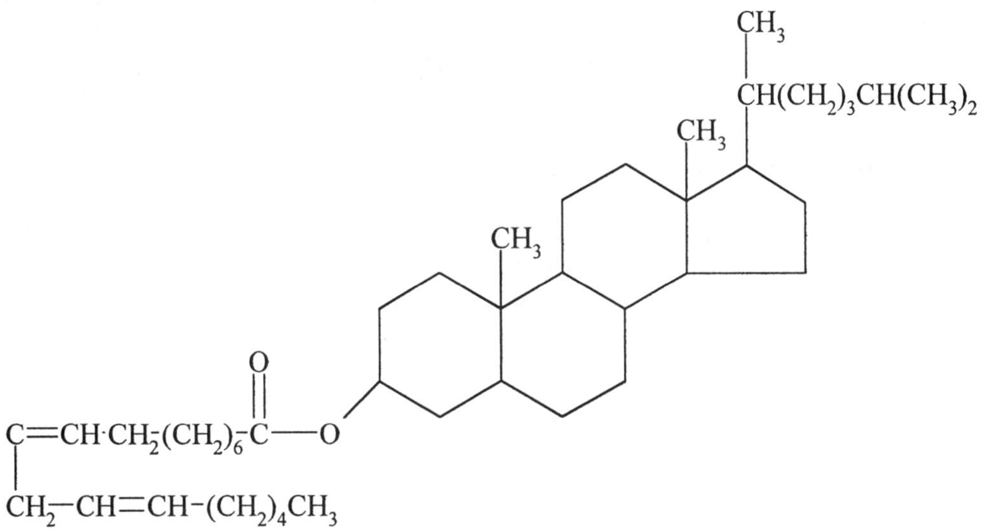

19.89 NaOH will react with the triacylglycerols in the fats to form glycerol and sodium carboxylate salts. Glycerol is soluble in water due to hydrogen bonding between H_2O and the –OH groups, while sodium carboxylate salts are also soluble in water due to the ionic nature of the compound. This increased solubility in water allows the clog to be washed out.

19.91 In biological membranes on Earth, the lipids organize to create an interior that is hydrophobic and an exterior that is hydrophilic, due to the presence of water throughout the body. On Gibo, we can assume that heptane is present everywhere in the body, and thus the membrane must organize to create an interior that is hydrophilic and an exterior that is hydrophobic. This can be accomplished by organizing the same lipids in a reverse bilayer, so that the hydrophobic portion is on the outside of the cell and the hydrophilic part is on the inside.

19.93 They both have a large hydrophobic (hydrocarbon) portion and a small hydrophilic (ionic) portion.

19.95 Hexane is not present in plants and animals, a property of all lipids.

19.97 The addition of hydrogen deceases the double bond content, but the reaction conditions also allow isomerization of double bonds. In this isomerization, the p bond breaks and reforms, yielding a double bond that is more likely to be trans than cis, because trans double bonds are more stable.

19.99 Infrared spectroscopy will differentiate these two molecules because testosterone has a hydroxyl group, while progesterone does not. Thus, the IR curve of testosterone will have a peak in the 3200-3650 cm^{-1} region, while that of progesterone will not.

19.101 The base structure of the triacylglycerol contains three carboxyl linkages, which contains three carbons and six oxygens. Thus, the remaining hydrocarbon chains must have the formula $C_{46}H_{88}$. If these hydrocarbon chains contained no double bond, they would have the formula C_nH_{2n+2}, which would be $C_{46}H_{94}$. Each C=C double bond has two less hydrogens than a single bond, thus this triacylglycerol must have three carbon-carbon double bonds.

19.103 This fat or oil must consist of a mixture of different triacylglycerols.

19.105 A catalyst lowers the activation energy of a reaction to speed it up, and thus the E_a of the catalyzed reaction is lower than that of the uncatalyzed reaction. However, because the reactants and products are the same for the uncatalyzed and catalyzed reactions, the ΔH of these two reactions is the same.

Chapter 20

Proteins

20.1 An α–amino acid is one that has the nitrogen bonded to the carbon that is right next to the carbonyl, a β–amino acid is one that has the nitrogen bonded to the carbon that is two carbons from the carbonyl, and a γ–amino acid is one that has the nitrogen bonded to the carbon that is three carbons from the carbonyl. Thus

(a) β; (b) α; (c) γ

20.3 (a) Gly; (b) Ser, Thr; (c) Phe, Tyr, Trp; (d) Ile, Thr; (e) Trp, His; (f) Met, Cys; (g) Asp, Glu; (h) Pro

20.5 The carboxylic acid group in the amino acid reacts with the amino group in an acid base reaction to form a positively charged amino group and a negatively charged carboxyl group.

20.7 (a) At its isoelectric point, alanine exists as a zwitterion. At pH < 1, the solution is very acidic. Under these conditions, the negatively charged carboxyl group reacts with free protons, which results in an amino acid that has a net positive charge. At pH > 12, the solution is very basic where the positively charged amino group will give up a proton to leave an amino acid that has a net negative charge. Thus:

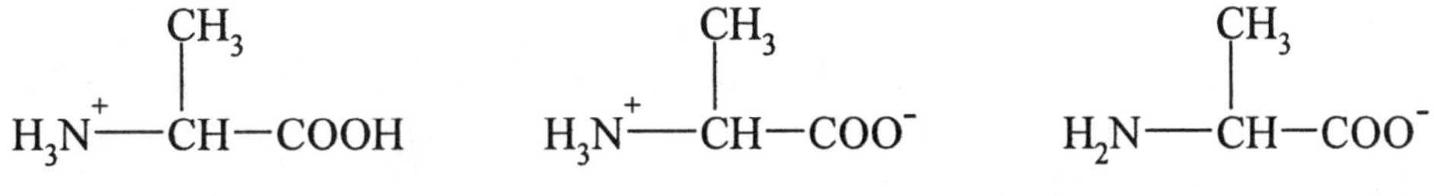

(b) Alanine is a neutral compound at pI, but is an ion at higher and lower pH. Ions are more soluble in water than neutral compounds, thus alanine is least soluble in water at pI = 6.01. It will be most soluble at very high and very low pH.

(c) At pI, the molecule is neutral and does not migrate in electrophoresis. At pH = 7, the pH is within 2 of pI, and thus will not migrate in electrophoresis. Alanine migrates to the anode at pH > 12, to the cathode at pH < 1.

20.9 • At pH = 7, the carboxylate and amino groups are charged.

• To name peptides, (i) the C-terminal residue keeps its amino acid name, (ii) the –ine or –ic acid of the other amino acid residues is replaced with –yl, and (iii) naming begins at the N-terminal residue.

Thus, Ser-Met is Serylmethionine:

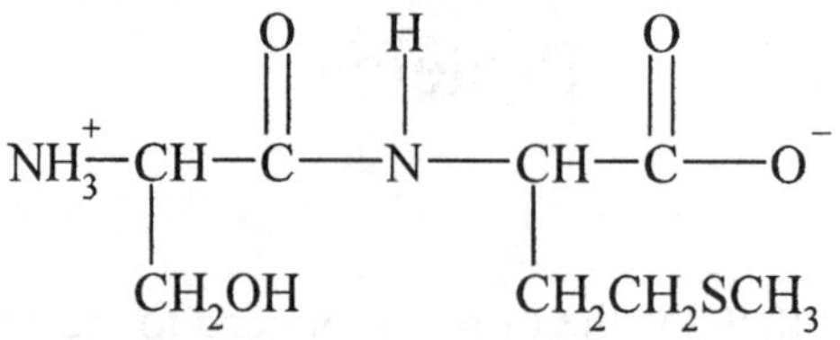

20.11 • At pH = 7, the carboxylate and amino groups are charged.

• To name peptides, (i) the C-terminal residue keeps its amino acid name, (ii) the –ine or –ic acid of the other amino acid residues is replaced with –yl, and (iii) naming begins at the N-terminal residue.

Thus, Thr-Ala-Asp is Threonylalanylaspartic acid:

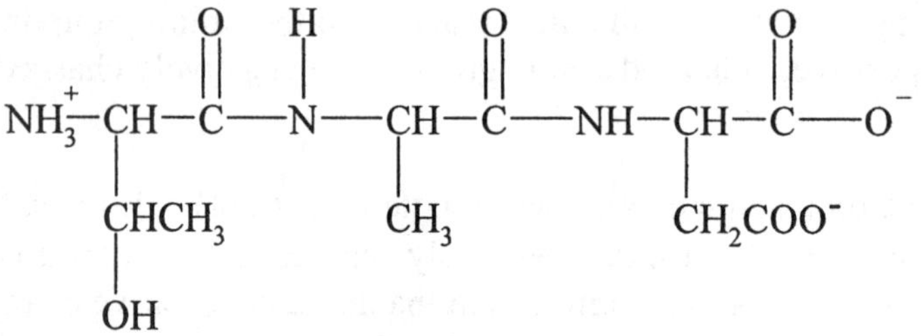

20.13 • At pH = 7, the carboxylate and amino groups are charged

• To name peptides, (i) The C-terminal residue keeps it's amino acid name, (ii) the –ine or –ic acid of the other amino acid residues is replaced with –yl and, (iii) naming begins at the N-terminal residue.

Thus, Gly-Pro-Ala is Glycylprolylalanine:

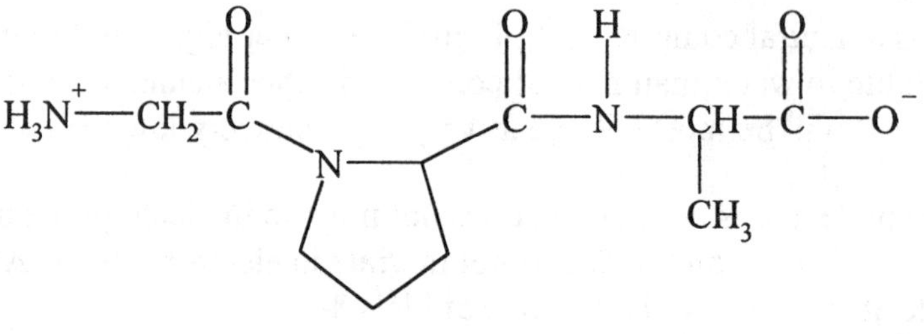

20.15 Constitutional isomers of a tripeptide exist as molecules with different connectivities of the three amino acids. For threonine, cysteine, and leucine there are six constitutional isomers: Thr-Leu-Cys, Thr-Cys-Leu, Cys-Leu-Thr, Cys-Thr-Leu, Leu-Cys-Thr, and Leu-Thr-Cys.

20.17 Each residue can exist as a D or L enantiomer. Thus, the six possible stereoisomers are different combinations of these enantiomers as: L-Lys-L-Ala-L-Glu, L-Lys-L-Ala- D-Glu, L-Lys-D-Ala-D-Glu, D-Lys-D-Ala-D-Glu, D-Lys-L-Ala-L-Glu, and D-Lys-D-Ala-L-Glu. Only the stereoisomer with all L-residues would be found in nature.

20.19 Aspartic acid has a carboxylate group on the side chain. Thus this tripeptide has three carboxylate groups and one amino group that are all charged at physiological pH. The tripeptide has a charge of 2- and will migrate to the positive anode.

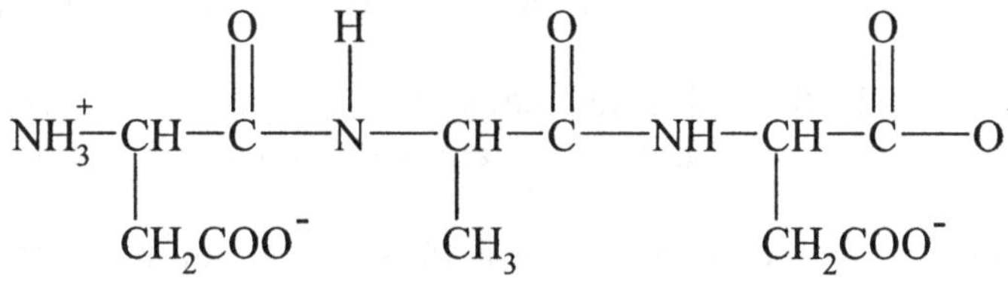

$$NH_3^+-CH-\overset{\overset{O}{\parallel}}{C}-\overset{\overset{H}{|}}{N}-CH-\overset{\overset{O}{\parallel}}{C}-NH-CH-\overset{\overset{O}{\parallel}}{C}-O^-$$

with CH_2COO^-, CH_3, CH_2COO^- side chains

20.21 In digestion, acidic hydrolysis, and basic hydrolysis, the peptide bonds are broken to leave the individual amino acids, Phe, Asp, Thr, and Lys. However, in acidic hydrolysis, the acidic environment results in the amino groups of the resultant amino acids being charged while the carboxylate groups are uncharged. In basic hydrolysis, the basic environment results in the carboxylate groups of the resultant amino acids being charged while the amino groups are uncharged.

20.23 Hydrolysis breaks the peptide bonds to show the composition of the original tripeptide, it must consist of Ala, Gly, and Lys. The products, however, do not show how these three amino acids were bonded together, i.e., the sequence. Thus the original tripeptide may be Ala-Gly-Lys, Ala-Lys-Gly, Lys-Ala-Gly, Lys-Gly-Ala, Gly-Ala-Lys, or Gly-Lys-Ala.

20.25 (a) Digestion breaks the peptide bonds to leave:

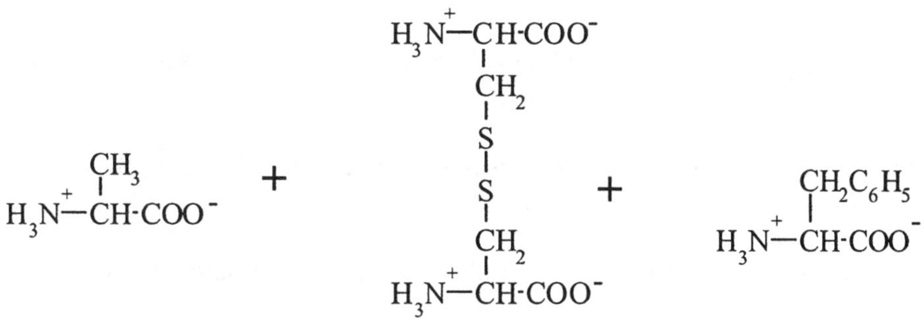

$$H_3N^+-CH\cdot COO^-$$
with CH_2—S—S—CH_2 chain

$$H_3N^+\overset{\overset{CH_3}{|}}{-}CH\cdot COO^- \quad + \quad \text{(cystine dimer)} \quad + \quad H_3N^+\overset{\overset{CH_2C_6H_5}{|}}{-}CH\cdot COO^-$$

(b) There are no functional groups in this hexapeptide that will selectively oxidize, thus there is no reaction.

(c) The disulfide bond can be reduced to form two tripeptides:

$$\underset{\substack{| \\ CH_2SH}}{H_3\overset{+}{N}-CH-CONH-\overset{\overset{\displaystyle CH_3}{|}}{CH}-CONH-\overset{\overset{\displaystyle CH_2C_6H_5}{|}}{CH}-COO^-}$$

20.27   Simple proteins contain only molecules that are made up of amino acids, while conjugated proteins contain nonpeptide molecules or ions as well as peptide molecules.

20.29   Conformation defines the three-dimensional structure of a large molecule. This three-dimensional structure can be changed by rotation about a single bond.

20.31   Hydrogen bonding between peptide groups is most important in determining the secondary structure of polypeptides.

20.33   (a) Proline is nonpolar, while Histidine is a basic amino acid, thus there is no attractive interaction between these two amino acids.

(b) Serine has an –OH group on the residue and tyrosine has an –OH on the residue, thus hydrogen bonds can form between these two amino acids.

(c) Both proline and phenylalanine have nonpolar side groups, thus hydrophobic interactions can occur between these two amino acids.

(d) Both lysine and arginine are basic amino acids, thus there is no attractive interaction between these two amino acids.

(e) Lysine is a basic amino acid while glutamic acid is an acidic amino acid, thus these two amino acids can undergo an acid-base reaction to form a salt bridge between the two amino acids.

(f) Serine is a polar amino acid while valine is a nonpolar amino acid, thus there is no attractive interaction between these two amino acids.

20.35   Nonpolar residues must be kept from water, hydrogen bonds can occur between peptide units, and polar residues can interact with water.

20.37   The primary structure defines the secondary, tertiary, and quaternary structure of a protein.

20.39   Fibrous proteins can aggregate by strong secondary attractive interactions to form macroscopic structures that are strong. This formation of strong macroscopic structures is imperative in the function of structural and contractile proteins.

20.41   A double helix is composed of two polypeptide chains that are entwined together, while a triple helix is composed of three polypeptide chains that are entwined together.

20.43 α-Keratins provide structural shape to cilia, hair, horns, hooves, skin, and wool.

20.45   When proteins function as catalysts, regulators, transporters, or protectors, they must not aggregate together to form macroscopic structures. The globular shape of globular proteins is not conducive to aggregation and thus they can fulfill this requirement. Their globular structure is also conducive to dissolution due to their hydrophilic surfaces.

20.47   Tertiary structure describes the relation between different conformational patterns within a given polypeptide. α-Keratins form a helix throughout the polypeptide chain and therefore do not possess different conformational patterns. Myoglobin, however, contains helical portions and β-sheet portion in different parts of the polypeptide chains. The arrangement of these helical and β-sheet structures in space is the tertiary structure of myoglobin.

20.49   The prosthetic group is the nonpeptide part of a conjugated protein, while the polypeptide is the apoprotein. Thus, the polypeptide is the apoprotein, while the heme is the prosthetic group.

20.51   Heme has an Fe^{2+} ion that bonds with the oxygen to hold it.

20.53   The amino acid residues that are on the exterior of globular proteins must be polar groups so that they can interact preferentially with water so that the protein can be dissolved. Thus Asp, Glu, His, and Lys will be on the exterior of the globular protein.

20.55   A mutation is an alteration in the DNA structure of a gene. This alteration sets off a chain of events, it will produce a change in the primary structure of a protein that is affected by the DNA, which in turn may alter the function of that protein.

20.57   Serine has a polar residue, while both phenylaniline and isoleucine have nonpolar residues. Thus the replacement of Phe with Ser alters the polar properties of that portion of the peptide chain; however, Ile replaces Phe, that part of the peptide still remains nonpolar.

20.59   The part of the peptide where this mutation occurs will become more polar. The peptide will thus attempt to move this portion to the exterior of the protein, which results in a change in the three-dimensional structure of the protein. This in turn will alter its function.

20.61   A native protein is one that has the conformation that exists under physiological conditions.

20.63   Digestion breaks the peptide linkages in a protein to alter its primary structure and produce amino acids. Denaturation alters the interactions between residues that give the secondary, tertiary, and quaternary structures of a protein. Thus denaturation changes the secondary, tertiary, and quaternary structures.

20.65 Ag^+ reacts with sulfhydryl groups of cysteine residues within the gonorrhea organism to metal disulfide bridges. These metal disulfide bridges alter the conformation of a protein, which in turn alters its function. This results in the death of the gonorrhea organism.

20.67   A change in pH will alter the charged state of the basic and acidic groups of the amino acid residues. Thus, a change in pH will alter the salt bridges that are formed from the charged acidic and basic groups in the residues. If a peptide does not have many acidic and basic residues, pH will not alter the structure.

20.69   To denote D and L designation, the structure must be drawn such that –COOH and –CH$_2$CH$_2$COOH are vertical in the structure. In the L structure, NH$_2$ is on the left. Thus, the structure is:

$$\begin{array}{c} \text{COOH} \\ | \\ \text{H}_2\text{N}\!-\!\!-\!\!-\!\text{H} \\ | \\ \text{CH}_2\text{CH}_2\text{COOH} \end{array}$$

20.71   (a) At its isoelectric point, methionine exists as a zwitterion. At pH < 1, the solution is very acidic. Under these conditions, the negatively charged carboxyl group reacts with free protons, which results in an amino acid that has a net positive charge. At pH > 12, the solution is very basic where the positively charged amino group will give up a proton to leave an amino acid that has a net negative charge. Thus:

pH < 1   pH ~ pI and 7   pH > 12

$$\begin{array}{ccc}
\text{CH}_2\text{CH}_2\text{SCH}_3 & \text{CH}_2\text{CH}_2\text{SCH}_3 & \text{CH}_2\text{CH}_2\text{SCH}_3 \\
| & | & | \\
\text{H}_3\overset{+}{\text{N}}\!-\!\text{CH}\!-\!\text{COOH} & \text{H}_3\overset{+}{\text{N}}\!-\!\text{CH}\!-\!\text{COO}^- & \text{H}_2\text{N}\!-\!\text{CH}\!-\!\text{COO}^-
\end{array}$$

(b) Methionine is a neutral compound at pI, but is an ion at higher and lower pH. Ions are more soluble in water than neutral compounds, thus methionine is least soluble in water at pI. It will be most soluble at very high and very low pH.

(c) At pI, the molecule is neutral and does not migrate in electrophoresis. At pH = 7, the pH is within 2 of pI, and thus will not migrate in electrophoresis. At pH = 1, methionine is positively charged and thus will migrate to the negatively charged cathode. At pH = 12, the amino acid is negatively charged and will migrate to the positively charged anode under these conditions.

20.73   There are 5! different constitutional isomers = 120. The two isomers that have Val as the N-terminal, proline as the C-terminal, and lysine as the middle residue are Val-Phe-Lys-Thr-Pro and Val-Thr-Lys-Phe-Pro.

20.75   Glycine has two –H's bonded to the carbon and thus does not have a tetrahedral stereocenter. Therefore in this tripeptide, only Lys and Ala can attain either L or D configuration. The four stereoisomers are thus L-Lys-Gly-L-Ala, D-Lys-Gly-D-Ala, D-

Lys-Gly-L-Ala, and L-Lys-Gly-D-Ala. Only the isomer with all L configuration amino acids will be found in nature, L-Lys-Gly-L-Ala.

20.77   • At pH = 7, the carboxylate and amino groups are charged.

• To name peptides, (i) The C-terminal residue keeps its amino acid name, (ii) the –ine or –ic acid of the other amino acid residues is replaced with –yl, and (iii) naming begins at the N-terminal residue.

Thus, Met-Ser-Ile-Gly-Glu is Methionylserylisoleucylglutamic acid and has a structure of:

$$
\begin{array}{ccccc}
& & CH_2CH_3 & & \\
& & | & & \\
CH_2CH_2SCH_3 & & CHCH_3 & & CH_2CH_2COO^- \\
| & & | & & | \\
H_3N^+\!-\!CH\text{-}CONH\text{-}CH\text{-}CONH\text{-}CH\text{-}CONH\text{-}CH_2\text{-}CONH\text{-}CH\text{-}COO^- \\
| & & & & \\
CH_2OH & & & &
\end{array}
$$

20.79   At pH = 7, the carboxylate and amino groups are charged. Therefore its structure is:

$$
\begin{array}{ccccc}
& O & H & O & O \\
& \| & | & \| & \| \\
NH_3^+\!CH\text{-}C\!-\!N\text{-}CH\text{-}C\!-\!NH\text{-}CH\text{-}C\!-\!O^- \\
| & & | & & | \\
CH_2COO^- & & (CH_2)_4NH_3^+ & & (CH_2)_4NH_3^+
\end{array}
$$

There is one more amino group than carboxylate, thus it has an overall charge of +1 and migrates to the negatively charged cathode. The pI of this compound will be slightly more basic than that of a neutral background to account for the +1 charge.

20.81   Phe-Asp-Leu-Glu-Thr has two more carboxylate groups (from Glu and Asp) than amino groups and will therefore have a –2 charge.

20.83   When this peptide is reduced, the disulfide bridge will be broken to form the peptide:

Gly-Ala-Cys-Gly-Lys-Phe-His-Glu-His-Cys-Met-Gly-Asp

20.85   (a) Albumin is a neutral compound at pI, but is an ion at higher and lower pH. Ions are more soluble in water than neutral compounds, thus albumin is least soluble in water at pI = 4.9.

(b) At physiological pH, the compound is more basic than at its pI and is therefore negatively charged. Therefore, it will migrate to the positively charged anode.

20.87   The compound must be negatively charged at physiological pH to migrate to the positively charged anode. Therefore, the compound must be in a more basic environment than its pI. Therefore, the compound must be pepsin.

20.89   Globular proteins require water solubility while fibrous proteins do not, thus globular proteins must have more polar residues on their surface.

20.91   X cannot be a β-pleated sheet because it has a low content of Gly, Ala, and Ser. It cannot be an α-helix either as it contains significant amounts of the helix breakers proline and glutamic acid. Therefore X is either a β-turn or a loop.

20.93   (a) Trp is a very large residue, while Gly is a very small residue. This replacement of a large residue with a small residue will alter the three-dimensional structure of the protein and thus its function.

(b) Both Arg and Lys are similar-sized basic residues and thus their exchange does not alter the structure of the protein.

(c) Both Thr and Ser are similar sized neutral polar residues and thus their exchange does not alter the three-dimensional structure and function of the protein.

(d) Arg is a basic residue, while Asp is an acidic residue. This exchange for acid and base will alter salt bridge interactions that control the three-dimensional structure. This, in turn, will alter the function of the protein.

20.95   Residue 1 must be formed from lysine by addition of an –OH group to the first carbon in the side group. Residue 2 must be formed from proline by addition of an –OH group to the cyclopentane ring. Residue 3 must be formed from glutamic acid by addition of another acidic group to the side chain.

20.97   (a) The interior of the bilayer is hydrophobic and thus will consist of nonpolar residues such as Leu, Phe, and Val.

(b) The extracellular environment is aqueous and thus must consist of hydrophilic residues such as Glu, Lys, and Ser.

(c) The intracellular environment is aqueous and thus must consist of hydrophilic residues such as Glu, Lys, and Ser.

20.99   The alteration in sickle-cell anemia hemoglobin occurs at a residue that strongly influences the conformation and function of the protein. The difference in residues between beef and human insulin must occur in regions that do not significantly alter the conformation and function of the protein.

20.101   Proteins are formed from many residues and are much larger molecules than the lipids.

20.103   Production of animal protein (such as cattle and poultry) is more expensive than production of vegetable protein (such as rice and beans).

20.105 New hair grows with the person's genetically determined straight shape.

20.107 20 residues × 3.5 Å/residue = 70 Å.

20.109 Insulin is a polypeptide that will breakdown in the digestive tract by hydrolysis. Thus, it will not reach the bloodstream as the whole molecule, which is required for its proper function.

20.111 The lower pI values indicate that there exists more acidic amino acid residues, thus the ratio of acidic-to-basic amino acid residues follows pepsin > albumin > lysozyme.

20.113 (a) At pH = pI, a molecule will not migrate in electrophoresis, thus 4.9.

(b) When pH > pI, albumin will have a negative charge.

(c) There is not pH when acidic and basic groups are uncharged; the response of the molecule to an electric field depends on the ratio of these charges.

20.115 Cysteine promotes more crosslinking between molecules, which creates a harder material. Thus the tortoise shell, which is harder, has more cysteine content.

Chapter 21

Nucleic Acids

21.1 Nucleotides are composed of a five-carbon sugar, phosphoric acid, and a heterocyclic nitrogen base. The five-carbon sugar in ribonucleotides is ribose while the five-carbon sugar in deoxyribonucleotides is deoxyribose.

21.3 In nucleotides, the phosphoric acid and sugar are bonded together. Thus, the nucleotide that is composed of phosphoric acid, deoxyribose, and thymine is:

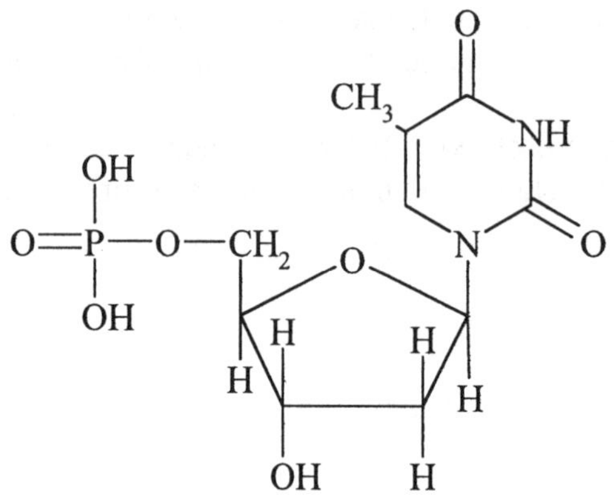

The abbreviated name consists of a "d" to denote deoxyribose, followed by the one-letter abbreviation for the base followed by "MP", which denotes 5'-monophosphate. Thus, the abbreviated name is dTMP.

21.5 dGMP is composed of guanine ("G"), deoxyribose ("d"), and phosphoric acid ("MP"). Hydrolysis breaks the bonds that were created in the formation of the nucleotide and thus will result in the products: deoxyribose, phosphoric acid, and guanine.

21.7 DNA is only found in the nucleus of the cell because it is too big to escape.

21.9 This compound is two nucleotides bonded together. The top one consists of phosphoric acid, deoxyribose, and thymine (dTMP) while the second consists of phosphoric acid, deoxyribose, and cytosine (dCMP). Therefore, this compound is dTMP-dCMP.

21.11 dUMP-dGMP contains deoxyribose and therefore would be a part of DNA. However, DNA does not contain uracil (U), therefore, this sequence cannot exist in nature.

21.13 This diad consists of a nucleotide that is in the 5' position that has guanine that is bonded to a nucleotide that contains thymine.

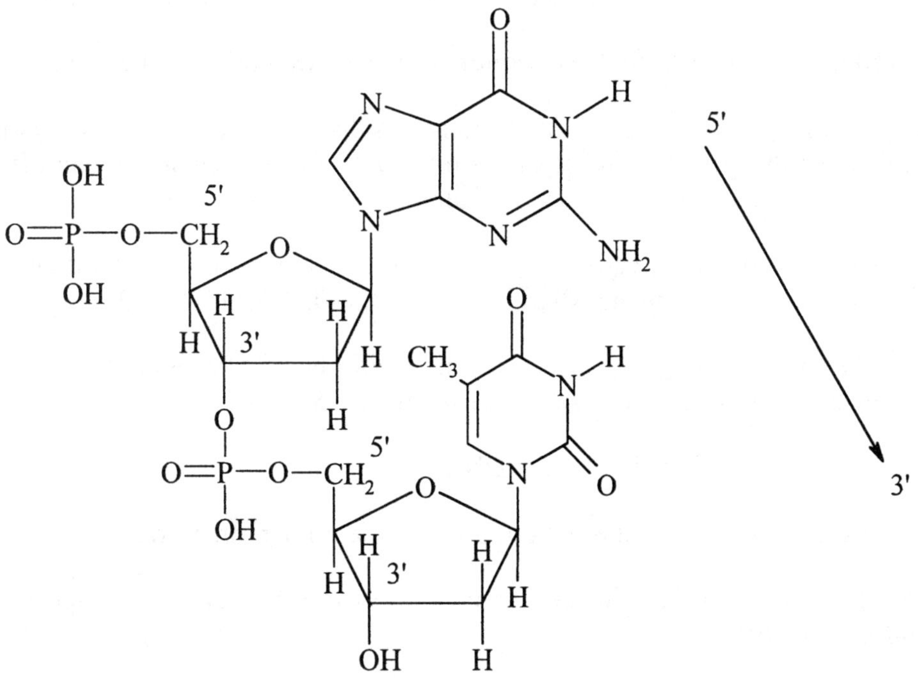

21.15 Hydrolysis breaks the linkage between nucleotides, thus it will produce dAMP, dGMP, dCMP, and two dTMP molecules.

21.17 Histones are proteins. Their function is to compress the large DNA double-helix molecule into the cell nucleus.

21.19 Base-pairing is the result of intermolecular hydrogen bonding.

21.21 (d). It cannot be (a), because A and T are always present in 1:1 ratio; it cannot be (b), the ratio of sugars to bases is 1:1; it cannot be (c) because in a given nucleic acid, all of the sugars are the same.

21.23 Adenine and thymine exist in a 1:1 ratio, while cytosine and guanine also exist in a 1:1 ratio. Therefore, there must exist 30.4 mol % thymine and 19.6 mol % cytosine.

21.25 The complementary base pair for A is T, for G is C, for C is G, and for T is A. Therefore, the complementary strand is 5'-CGGTTA-3'.

21.27 The sequence of bases along the DNA chain contains genetic information.

21.29 Replication produces a chromosome.

21.31 (a) Replication bubbles are the portion of the DNA double helix that is unwound and is the part of the helix where replication takes place.

(b) A leading strand is the DNA strand that is synthesized in a continuous manner; a lagging strand is a strand that is synthesized in a discontinuous manner.

(c) Okazaki fragments are the discontinuous DNA fragments of the lagging strand.

21.33   A parent DNA consist of two DNA molecules that form a double helix. During replication this parent strand separates into two DNA molecules that pair off with two new DNA strands to create a pair of daughter DNA molecules.

21.35   (a) The corresponding sequence in the old DNA chain of the daughter double helix consists of complimentary base-pairs and thus is 5'-CGCTAA-3'.

(b) The corresponding sequence in the old DNA chain of the other daughter double helix must consist of the same sequence and thus is 5'-TTAGCG-3'.

21.37   Transcription produces RNA from DNA.

21.39   The template DNA strand is used to produce the primary transcript.

21.41   Posttrascriptional processing occurs to modify various primary transcripts to form rRNA, mRNA, and tRNA.

21.43   DNA is orders of magnitude larger than any RNA; DNA typically has a molecular mass of many billions; rRNA typically has a molecular mass of 500,000 – 1,000,000; mRNA typically has a molecular mass between 100,000 and 1–2 million; tRNA typically has a molecular mass less than 50,000.

21.45   In RNA, U forms a base pair with A and C forms a base pair with G. Therefore, the composition of RNA formed from this sequence is 15 mol % U, 25 mol % G, 20 mol % C, and 40 mol % A.

21.47   Translation is the synthesis of polypeptides.

21.49   rRNA are a component that forms the ribosomes in which polypeptide synthesis takes place. mRNA carries the genetic message that encodes the amino acid sequence of the polypeptide that is synthesized. mRNA also binds to the ribosome and acts as the template for polypeptide synthesis. tRNA transports α-amino acids into the ribosome.

21.51   The sequence of bases in mRNA is the genetic message that it carries. This information specifies the sequence of α-amino acids in the polpeptide that is synthesized.

21.53   The genetic code is considered universal because almost every living species, including plants and animals, follows the same genetic code.

21.55   Aminoacyl-tRNA synthetase is the enzyme that catalyzes the concatenation of amino acids at aminoacyl and peptidyl sites on mRNA.

21.57   A base triplet is the name given to three consecutive bases along a DNA or RNA chain.

21.59   The primary transcript is merely a chain with base pairs that are complementary in RNA to the template DNA strand, thus the primary transcript is 3'-UGU-GUG-GUU-UAC-ACA-CCA-GUA-5'.

The mRNA is then formed only from the exons of the template DNA strand. It is formed from the complementary base pairs, and is therefore, 5'-AUG-CAU-UUG-UGU-3'.

This mRNA sequence codes to the polypeptide Met-His-Leu-Cys.

21.61   A mutation is an error in the base sequence of a gene.

21.63   In a substitution mutation, the identity of a base along a sequence is changed, while in a frameshift mutation a base is inserted or deleted into a sequence such that the sequence is shifted.

21.65   A spontaneous mutation is one that results from a random error in replication.

21.67   Germ cells are cells that contain information that can be passed onto offspring or future generations. Somatic cells are all other cells.

21.69   A cancer is an uncontrolled growth and division of cells to form tumors.

21.71   Physiological function is most often affected by a mutation when a residue is replaced by another that is of different size, charge, or polarity. In mutation (a), a negatively charged amino acid residue is replaced by an uncharged residue, while in mutation (b), one negatively charged residue is replaced by another. Thus, mutation (a) is more harmful.

21.73   The peptide that is synthesized is determined by the sequence of base pairs in each codon that results from the exons. This procedure begins at the 3' end. Thus, the first exon is TAC which correlates the codon AUG (T pairs with A, A pairs with U, and C pairs with G). Thus, the following two exons (CTG and AAT) result in the codons GAC and UUA.

(a) Using Table 21.1, AUG codes for the synthesis of Met, GAC codes for Asp, and UUA codes for Leu. Thus, Met-Asp-Leu will be synthesized from this DNA strand.

(b) If GTC triplet is mutated to CTC, this exon will produce the codon GAG that will result in the synthesis of Glu rather than Asp, thus the polypeptide Met-Glu-Leu will be synthesized.

(c) If GTC triplet is mutated to ATC, this exon will produce the codon GAU that will result in the synthesis of Asp, thus the polypeptide Met-Asp-Leu will still be synthesized.

(d) There is no effect of a mutation in the intron on the process exon → codon → polypeptide, thus the polypeptide Met-Asp-Leu will still be synthesized.

21.75   An antibiotic is a chemical, which is usually an organic compound that kills microorganisms such as bacterium, mold, or yeast.

21.77   A virus is an infectious, parasitic particle that consists of either DNA or RNA, but not both, and is encapsulated in a protein coat. A virus is also usually smaller than bacteria.

21.79   A vaccine contains a weakened virus or its proteins. Its purpose is to prevent viral infections by stimulating the immune system to generate antibodies that will recognize the virus.

21.81   Recombinant DNA technology relies on the principle that transplanting DNA from one organism into another can alter the genome of the other organism.

21.83   In recombinant DNA technology, the vector DNA is the organism's DNA that is to be altered. The donor DNA is the DNA that is introduced into the vector DNA to create this alteration. The altered DNA is the recombinant DNA.

21.85   Bonding many nucleotides together forms nucleic acids. Thus, nucleotides are building blocks for nucleic acids.

21.87   The 5'-end of a nucleic acid has only one of the oxygens of the phosphate group at C5 attached to a carbon; the 3'-end of the nucleic acid has an unreacted –OH at C3 of the sugar.

21.89   This abbreviated description describes a sequence with deoxyribose and is thus part of DNA. However, DNA is formed from C, T, A, and G. There are no deoxyribonucleic acids that contain U.

21.91   Hydrolysis breaks the linkage between nucleotides, thus it will produce dAMP, dCMP, dGMP, and dTMP in a ratio of 1:1:2:3.

21.93   DNA exists in the form of a double helix consisting of two DNA strands, while RNA is single stranded.

21.95   Adenine forms a base pair with thymine while cytosine forms a base pair with guanine. Thus for every adenine in one strand, there exists the same amount of thymine in the other strand. The same relationship holds for cytosine and guanine between the two strands.

(a) 14 mol% A and 31 mol % C

(b) 17 mol % C and 38 mol % A

21.97   The primary transcript is a chain with base pairs that are complementary in RNA to the template DNA strand, thus the primary transcript is 3'-AUG-ACC-ACA-CAA-UUU-GUG-AGU-5'.

The mRNA is then formed only from the exons of the template DNA strand. It is formed from the complementary base pairs, and is therefore 5'-AUG-CAA-UUU-AGU-3'.

This mRNA sequence codes to the polypeptide Met-Glu-Phe-Ser.

21.99   A transgenic plant or animal is one that is conceived from a germ cell containing recombinant DNA.

21.101 Gene therapy seeks to correct hereditary diseases by introducing recombinant DNA into an organism.

21.103 Denaturation occurs when the hydrogen bonds that stabilize base pairs are broken. G-C pairs are stabilized by three hydrogen bonds, whereas A-T base pairs have two. This means that it is easier to break A-T pairs than G-C pairs. Therefore, the more G-C pairs that exist, the more thermal energy (and higher temperature) is needed to break this stronger interaction.

21.105 DNA consists of pairs of DNA molecules held together by the specific base pairs G-C and A-T. This specific pairing results in the number of G being equal to the number of C and the number of A being equal to the number of T in a given DNA double helix. RNA, however, exists as single molecules and thus does not require these specific ratios.

21.107 The synthesis of each amino acid residue is coded by a three-nucleotide base sequence, thus 146 amino acids will require at least 438 (= 3 × 146) nucleotide bases. There will usually be many more nucleotide bases because only the exons of the gene code to the synthesis of the polypeptide. There usually also exist introns that do not correlate to polypeptide synthesis.

21.109 The tendency of spontaneous mutations that produce antibiotic-resistant strains of the bacteria increases with increased use of the antibiotic.

21.111 $\dfrac{3.2 \text{ billion base pairs}}{46 \text{ chromosomes}} \times \dfrac{340 \text{ g}}{\text{mole}} = 23.7$ billion grams/mole

21.113 A two-base code system based on five different bases would code a total of 25 (5^2) different items, sufficient to code for the 20 different amino acids and the initiation and stop signals. Such a genetic code, however, would have very little redundancy and protection against mutation.

21.115 Proto-oncogenes are genes that are responsible for producing the regulatory proteins responsible for normal cell growth and function. Mutation of proto-oncogenes to oncogenes results in cancer because oncogenes are genes that no longer code for the correct regulatory proteins.

21.117 A reverse transcriptase inhibitor blocks the transcription of the HIV RNA to HIV DNA. A protease inhibitor blocks the cutting up of the proteins produced by translation of HIV RNA. A fusion inhibitor interferes with the attachment of the virus to the host-cell membrane.

Chapter 22

Enzymes and Metabolism

22.1 Biochemistry is the extension of the relation of the structure and function of cells to their organization on a molecular and subcellular level.

22.3 Eukaryotic cells contain subcellular membrane-bounded organelles or structures, while prokaryotic cells do not.

22.5 Oxidative processes take place within the mitochondria of animal cells.

22.7 Protein synthesis takes place at ribosomes.

22.9 DNA in bacterial cells is not bounded by a membrane and is located in a microscopically visible nuclear zone.

22.11 (i) To obtain energy in a chemical form by the degradation of nutrients.

(ii) To convert nutrient molecules into precursor molecules used to build cell macromolecules.

(iii) To synthesize cell macromolecules.

(iv) To produce or modify the biomolecules necessary for specific functions in specialized cells.

22.13 Anabolism is the group of processes that synthesize biomolecules.

22.15 The first stage of catabolism reduces nutrient macromolecules to monosaccharides, amino acids, and fatty acids.

22.17 Acetyl-S-CoA is oxidized to carbon dioxide, CO_2, and water, H_2O.

22.19 ATP functions as the carrier of energy to the energy-requiring processes of cells and is the link between catabolism and anabolism.

22.21 ATP can be synthesized by substrate phosphorylation, in which a phosphate group is added to ADP from a high-energy compound and oxidative phosphorylation, in which ADP reacts directly with an inorganic phosphate.

22.23 NAD^+ functions in catabolic reactions, while the reduced form of $NADP^+$ is primarily used in reductive biosynthetic reactions.

22.25 An active site is a region on an enzyme where catalysis takes place. It is a binding site and it has catalytic function.

22.27   Inducible fit describes cases in which the structure of a substrate molecule can influence the complementarity of a binding site.

22.29   Many vitamins become cofactors or serve as sources that can be transformed into cofactors.

22.31   The turnover number is the number of moles of substrate that react per unit time per mole of enzyme.

22.33   A regulatory enzyme is one whose activity can be modified by combination with specific activators or inhibitors.

22.35   High-energy compounds are responsible for driving forward chemical reactions that are unfavorable.

22.37   Cells handle the different types of molecules by breaking down the incoming compounds and dealing with the products in similar ways. Catabolism first collects functional molecules by degrading nutrient macromolecules to their building blocks, proteins to amino acids, and polysaccharides and hydrolyzable lipids to fatty acids and glycerol. After that, hexoses, the carbon chains of fatty acids, and most of the amino acids are converted to acetyl-S-CoA, which is eventually oxidized to carbon dioxide and water.

22.39   No. The activities of cellular enzymes are regulated so that they can respond appropriately to the immediate metabolic needs of the cell. This regulation can occur by various methods including the binding of molecules that are not their substrates to allosteric regulatory enzymes, by covalent modification, or by feedback inhibition.

22.41   Yes. A cofactor may be required for achieving an enzymes correct conformational structural requirements for substrate binding, and a separate coenzyme may be required for the catalytic process.

22.43   The feedback inhibitor is product E. Its concentration is controlled by its acting as an inhibitor of enzyme 2.

22.45   Competitive inhibition is reversed by increasing the concentration of substrate. Non-competitive inhibition cannot be reversed by increasing substrate concentration.

22.47   No. Example: Inhibition by heavy metals such as Ag^+, Hg^{++}.

22.49   False. Enzymes are catalysts. Catalysts enter into a chemical process in which reactants undergo changes to form products and emerge unchanged. They have no influence on the final changes of the chemical reaction.

Chapter 23

Carbohydrate Metabolism

23.1   Glucose is the principal source of energy for metabolism.

23.3   The most important products of glycolysis are ATP, NADH, and lactate (pyruvate under aerobic conditions).

23.5   The activity of the enzyme phosphofructokinase decreases at high concentrations of ATP. Therefore, this enzyme is the key regulatory control point in glycolysis.

23.7   In glycolysis, ATP is synthesized in two reactions, both of which occur during the second stage of glycolysis:

ADP + 3-bisphosphoglycerate $\rightarrow$ ATP + 3-phosphoglycerate

ADP + phosphoenolpyruvate $\rightarrow$ ATP + pyruvate

23.9   In active skeletal muscles, NAD^+ is regenerated by the reduction of pyruvate to lactate.

NADH + H^+ + pyruvate $\rightarrow$ NAD^+ + lactate

23.11   Glycolysis would cease if inorganic phosphate were unavailable.

23.13   The reaction of pyruvate with NAD^+ and CoA-SH to form acetyl-S-CoA occurs in the mitochondria.

23.15   Yes. It is deactivated by phosphorylation and activated by dephosphorylation.

23.17   Carbon dioxide is formed in two reactions of the citric acid cycle, when isocitrate reacts with NAD^+ to form α-ketoglutarate

Isocitrate + NAD^+ $\rightarrow$ α-ketoglutarate + CO_2 + NADH

And when α-ketoglutarate reacts with CoA-SH to form succinyl-S-CoA

α-ketoglutarate + CoA-SH $\rightarrow$ succinyl-S-CoA + CO_2

23.19   Because the products of the sequence of reactions form a circle. In other words, the product of the sequence, oxaloacetate, is also the first reactant of the same sequence.

23.21   Because oxaloacetate is regenerated at the conclusion of each cycle, one molecule of it can affect the oxidation of a limitless numbers of acetyl groups. Therefore, oxaloacetate acts as a catalyst for this cycle.

23.23   The concentration of oxaloacetate increases when the enzyme pyruvate carboxylase is activated by acetyl-S-CoA and, in the presence of ATP, catalyzes the formation of oxaloacetate from pyruvate and CO_2.

23.25   Not directly. Oxaloacetate is first synthesized in the mitochondria and reduced to malate, which then enters the cytosol and is reoxidized to oxaloacetate. There the oxaloacetate reacts with GTP under the influence of the enzyme phosphoenolpyruvate carboxykinase to form phosphoenolpyruvate, CO_2, and GDP.

23.27   Oxaloacetate is first reduced to malate, which can pass through the mitochondrial membrane and enter the cytosol.

23.29   Glycogen is the polymeric storage form of glucose in animal tissue.

23.31   The hydrolysis of pyrophosphate has a very large equilibrium constant, is coupled to the overall metabolic process, and is the driving force for the formation of the activated glucose molecule.

23.33   In phosphorolysis a glucose molecule is removed from glycogen by the formation of glucose-1-PO_4, whereas hydrolysis yields glucose.

23.35   Glucagon and epinephrine are hormones that initiate glycogenolysis.

23.37   They both regulate the blood-glucose concentration. Epinephrine also affects blood pressure and heart rate.

23.39   Only liver cells possess the enzyme glucose-6-phosphate phosphatase, which hydrolyzes glucose-6-PO_4 to form free glucose.

23.41   There are three locations along the electron-transport chain where there is sufficient energy for the synthesis of ATP.

23.43   No. The inner membrane of mitochondria must be intact for ATP synthesis to take place.

23.45   No. The inner membrane is impermeable to most compounds. Transport across this membrane occurs via transport proteins that are specific for only a few substances.

23.47   The maximum total amount of ATP produced under aerobic conditions is 38 mol of ATP per mole of glucose.

23.49   In active muscle cells, the rate of glycolysis is much greater than the rate of the citric acid cycle.

23.51   Active muscle cells produce ATP primarily through glycolysis. For glycolysis to continue at a maximal rate, NAD^+ must be regenerated by oxidation of NADH and inorganic phosphate must be available for the formation of 1,3-bisphosphoglycerate. Therefore, for muscle cells to produce ATP, NAD^+ must be regenerated quickly and inorganic phosphate must be available.

23.53 NADH delivers its reducing power to the electron-transport chain by reducing oxidized cytosolic substances to their reduced counterparts, which can then penetrate the mitochondrial membrane.

23.55 ATP is produced by substrate phosphorylation in glycolysis, and by oxidative phosphorylation in the citric acid cycle.

23.57 Both the rate of glycogenolysis and glycogenesis will be markedly reduced.

23.59 The conversion of pyruvate to acetyl-S-CoA.

23.61 Assuming seven residues in each branch, and branch formation every 11 residues, there will be 1000 branches each with a reducing end, hence 1000 reducing ends.

Chapter 24

Fatty Acid Metabolism

24.1 No. The upper limit for the mass of glycogen stored in the liver and muscles will only last about 12 hours.

24.3 When the blood-glucose concentration reaches its lowest levels between meals, glucagon is released.

24.5 The fatty acid is oxidized only in the form of an acyl-S-CoA thioester, therefore, the reaction that activates the fatty acid to oxidation is this transformation,

$$R\text{-}COO^- + ATP + CoA\text{-}SH \rightarrow RCO\text{-}S\text{-}CoA + AMP + 2\,P_i$$

24.7 No. There is no specific transport system for fatty acyl-s-CoA derivatives in the mitochondrial membrane.

24.9 The first step in fatty acid oxidation is a dehydrogenation between carbons 2 and 3. Thus, the product for a 16-carbon fatty acid is:

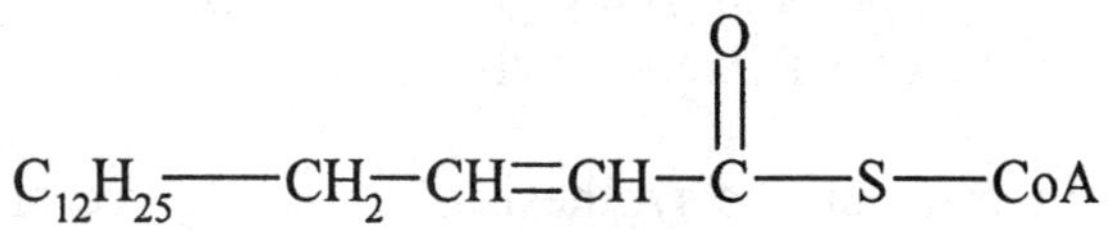

$$C_{12}H_{25}\text{---}CH_2\text{-}CH{=}CH\text{-}\overset{\displaystyle O}{\overset{\|}{C}}\text{---}S\text{---}CoA$$

24.11 The second step is the enzymatic hydrogenation of the trans double bond to form:

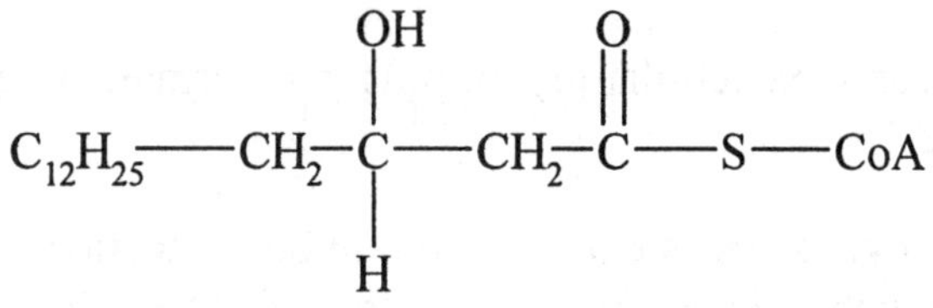

$$C_{12}H_{25}\text{---}CH_2\text{-}\underset{\underset{\displaystyle H}{|}}{\overset{\overset{\displaystyle OH}{|}}{C}}\text{---}CH_2\text{-}\overset{\displaystyle O}{\overset{\|}{C}}\text{---}S\text{---}CoA$$

24.13 The third step is a dehydrogenation to form:

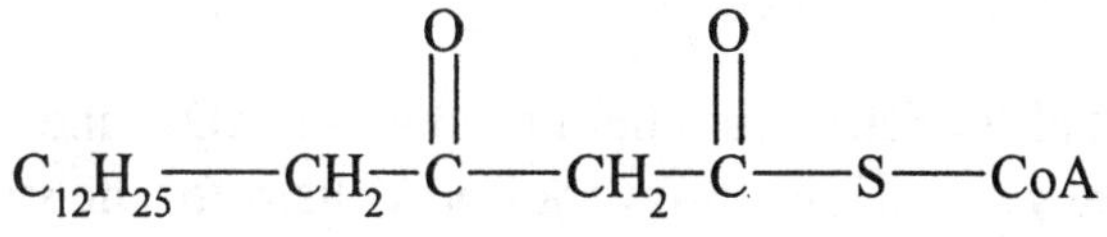

$$C_{12}H_{25}\text{---}CH_2\text{-}\overset{\displaystyle O}{\overset{\|}{C}}\text{---}CH_2\text{-}\overset{\displaystyle O}{\overset{\|}{C}}\text{---}S\text{---}CoA$$

24.15 The reactants in the fourth oxidation step are the product of the third step, 3-keto(C_{16})acyl-S-CoA, and CoA-SH.

24.17 The ketone bodies synthesized in the liver are acetoacetate, D-3-hydroxybutyrate, and acetone.

24.19 Ketosis occurs when acetyl-S-CoA is in excess, which usually occurs during starvation and in diabetes mellitus.

24.21 Fatty acid synthesis takes place in the cytosol, while oxidation or catabolism takes place in the mitochondria.

24.23 Insulin initiates fatty acid biosynthesis by activating the phosphodiesterase.

24.25 Citrate from the mitochondria serves as the principal source of acetate in fatty acid biosynthesis.

24.27 Acetyl-S-CoA is transferred to the α-SH site, and malonyl-S-CoA is transferred to the β-SH site.

24.29 CO_2 is lost as HCO_3^-, which makes the reaction irreversible and provides the driving force for the reaction. Biotin is the cofactor required for the formation of malonyl-S-CoA.

24.31 Prior to chain lengthening, the α-SH site is occupied by an acetyl group, and a carboxylated acyl derivative is at the β-site.

24.33 The first step in the biosynthesis of triacylglycerols begins with the formation of phosphatidate from glycerol-3-PO_4 and acyl-S-CoA.

24.35 Yes, it is worth the investment; the synthesis of a triacylglycerol requires about 15% of the ATP that is generated in its oxidation.

24.37 Ethanolamine is added to the diacylglycerol in the activated form of cytidine diphosphoethanolamine.

24.39 The hydrolysis of complex cellular glycolipids that normally takes place in the lysosomes does not take place.

24.41 In catabolism, the acyl carrier is CoA-SH, but in biosynthetic reactions, the acyl carrier is an -SH protein. Reduction in catabolism employs NADH, while in biosynthetic reactions, NADPH is used.

24.43 8 Acetyl-S-CoA + 7 ATP^{4-} + 14 NADPH + 7 H^+ $\rightarrow$ palmitoyl-S-CoA + 14 $NADP^+$ + 7 CoA-SH + 7 ADP^{3-} + 7 P_i

24.45 Yes. The oxidation of the fatty acid chain produces NADH and $FADH_2$, both of which provide reducing power to the electron-transport chain for the synthesis of ATP through oxidative phosphorylation.

24.47 True. The presence of high concentrations of citrate indicates that the cells are in a high-energy state (ATP in high concentration). Equally important is the fact that citrate is a specific allosteric activator of acetyl-S-CoA carboxylase, which catalyzes the rate-limiting step in the fatty acid synthase system.

24.49   Fatty acids arise in adipose tissue by enzymatic hydrolysis of the stored triacylglycerols. The hydrolysis is catalyzed by a lipase that is activated by glucagon. The fatty acids leave adipose cells, become solubilized by being bound to serum albumin, and in that form travel throughout the circulatory system.

24.51   It (1) allosterically activates acetyl CoA carboxylase and (2) provides a mechanism for transporting acetyl CoA from the mitochondria.

24.53   Sam doesn't realize that excess glucose produces excess acetyl CoA, which then triggers synthesis of fatty acids, stored as triglycerides. He must count calories if his aim is to keep weight down.

24.55   The absence of the lysozymal enzyme, N-acetylhexosaminidase, causes cellular lysozymes to swell out of control, leading to cell death.

24.57   When an individual has familial hypercholesterolemia.

Chapter 25

Amino Acid Metabolism

25.1 Transamination is a process in which the amino group of an amino acid is interchanged with the carbonyl group of an α-keto acid.

25.3 Excess amino acids not used in biosynthesis are catabolized and used as energy sources.

25.5 The first step in the catabolism of ingested amino acids is when the amino groups are transferred to α-ketoglutarate by transamination.

25.7 NH_4^+ from liver cells is rendered nontoxic by converting into urea.

25.9 A urotelic animal excretes ammonium ion in the form of urea.

25.11 If there is an excess concentration of NH_4^+ in cells, it reacts with a large amount of α-ketoglutarate. This results in the excessive lowering of α-ketoglutarate concentrations.

25.13 Muscle cells use the glucose-alanine cycle to render NH_4^+ harmless for transport through the blood.

25.15 The reactions constituting the urea cycle begin in the cytosol, continue in the mitochondria, and end in the cytosol.

25.17 Citrulline is an amino acid that participates in urea's synthesis.

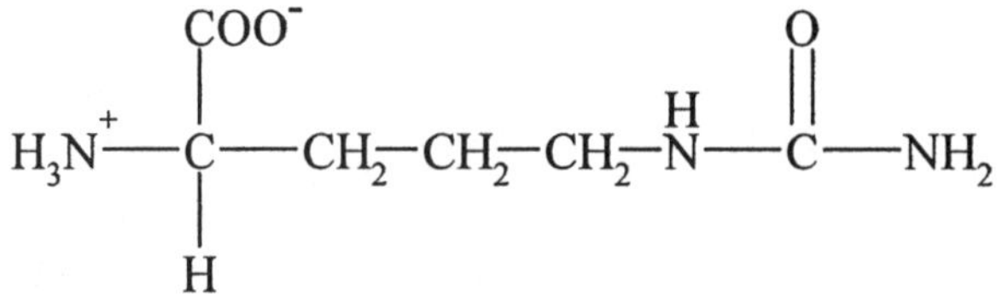

$$H_3N^+ - \overset{\overset{\displaystyle COO^-}{|}}{\underset{\underset{\displaystyle H}{|}}{C}} - CH_2 - CH_2 - CH_2 - \overset{\displaystyle H}{N} - \overset{\overset{\displaystyle O}{\|}}{C} - NH_2$$

25.19 ATP is not required for the formation of citrulline, the formation of arginine and fumarate from argininosuccinate, and the hydrolysis of arginine to form urea and ornithine.

25.21 Approximately 20% of the energy available in amino acids is used to synthesize urea.

25.23 The catabolism of these amino acids provides products that can be used to synthesize glucose.

25.25 The catabolism of these amino acids results in the formation of ketone bodies.

25.27 When phenylalanine is catabolized, acetoacetyl-S-CoA and fumarate are formed.

Fumarate can be used to create glucose through gluconeogenesis, while acetoacetyl-S-CoA yields ketone bodies in the liver. Therefore, the product of its catabolism results in the formation of both glucose and ketone bodies.

25.29   The result of the metabolic defect causes the intermediates of catabolic or anabolic sequences to accumulate in cells.

25.31   Phenylketonuria leads to severe mental retardation. Treatment requires that phenylalanine be eliminated or severely restricted from the diet of newborns.

25.33   A nonessential amino acid is one that can be synthesized by humans.

25.35   The nonionic water-soluble compound urea can be excreted without the loss of other important anions such as phosphate.

25.37   No. Arginase is an enzyme found only in the livers of terrestrial animals.

25.39   Phenylketonuria, which causes defects in the central nervous system, is the result of a defect in the enzyme that oxidizes phenylalanine to form tyrosine. The hydrolysis of aspartame in the intestine will produce phenylalanine, which phenylketonurics must avoid.

25.41   False. Oxidative deamination of aspartate leads to the formation of pyruvate and ammonium ion. Ammonia cannot exist at physiological pH.

25.43   C could inhibit the A → B step, Y could inhibit the C → D step, and Z could inhibit the C → G step.

Another possibility is: Y could inhibit the C → D step, and Z could inhibit the C → G step, and the A → B step would be inhibited only if Y and Z are both present.

25.45   False. The end products of leucine catabolism are acetyl-S-CoA, and acetoacetate. Neither of these compounds are components of gluconeogenesis.

Chapter 26

Nutrition, Nutrient Transport, and Metabolic Regulation

26.1 Foods are enzymatically degraded to low-molecular-mass components to prepare them for absorption in the gut.

26.3 Gastrin is a hormone stimulated by the entry of protein into the stomach

26.5 A zymogen is an inactive enzyme precursor.

26.7 They are transported across intestinal cells into the bloodstream and directly into the portal circulation.

26.9 Gastrin stimulates the secretion of pepsinogen and HCl.

26.11 If a given protein provides all the required amino acids in the proper proportions and all are released upon digestion and absorbed, the protein is said to have a biological value of 100.

26.13 Fatty acids containing more than one unsaturated bond past carbon 9 of a saturated chain, counting from the carboxyl end, cannot be synthesized by humans. Eicosanoids, a family of lipid-soluble organic acids that are regulators of hormones, are synthesized by mammals from arachidonic acid, a polyunsaturated fatty acid that mammals can synthesize with the use of dietary polyunsaturated fatty acids of plant origin as precursors.

26.15 A dietary deficiency disease is caused by a deficiency in a factor essential to cellular function that can only be obtained through the diet.

26.17 The cells present are erythrocytes, leukocytes, and platelets.

26.19 A lipoprotein consists of a core of hydrophobic lipids surrounded by a shell of amphipathic lipids and proteins.

26.21 The proteins in lipoproteins solubilize lipids, direct specific lipoproteins to particular tissues, and activate enzymes that hydrolyze and unload lipids from lipoproteins.

26.23 The presence of hemoglobin in eythrocytes.

26.25 It lowers the affinity of hemoglobin for oxygen.

26.27 Bicarbonate ion is formed in erythrocytes from CO_2, under the influence of the enzyme carbonic anhydrase.

26.29 A high concentration of CO_2 lowers the pH, and the Bohr effect enhances oxyhemoglobin dissociation.

26.31   The blood pH rises to higher than normal levels because CO_2 is being eliminated faster than it is being formed by respiring cells.

26.33   The brain uses mostly glucose and some 3-hydroxybutyrate for energy needs.

26.35   Virtually none.

26.37   It must depend on the citric acid cycle for its ATP.

26.39   Glycogen.

26.41   Lipases on adipocyte cell surfaces hydrolyze the triacylglycerols of the chylomicrons, allowing the resulting fatty acids to enter the cells.

26.43   Glucagon stimulates lipases within adipocytes.

26.45   It eventually reaches the liver to contribute to gluconeogenesis.

26.47   To produce the ATP necessary to effect the asymmetric distribution of Na^+ ions around the kidney tubules.

26.49   Excessive amounts of hydrogen ion can be eliminated by increasing the ammonia concentration of the urine, thereby increasing the amount of ammonium ion excreted.

26.51   The liver alone possesses glucokinase, an enzyme able to phosphorylate glucose at the high concentrations present in the blood after a meal.

26.53   It is converted into urea.

26.55   Acetoacetate and 3-hydroxybutyrate.

26.57   It synthesizes the bile acids from cholesterol.

26.59   No. For example, the hormone insulin is a protein.

26.61   No. Steroid hormones penetrate cell membranes and are transported to the cell nucleus to modify DNA translation.

26.63   No. The cellular uptake of oxygen by the brain remains constant under all conditions.

26.65   False. Neurotransmitters have a very short life.

26.67   The hypothalamus secretes a specific hypothalamic releasing hormone (HRH) that causes the anterior pituitary gland to secrete thyrotropic hormone.

26.69   A large electrical potential across its cell membranes and the ability to transmit an electrical signal from cell to cell.

26.71 $Na^+_{in} < 10$ mM, $Na^+_{out} = 150$ mM; $K^+_{in} = 125$ mM, $K^+_{out} < 5$ mM

26.73   The brain synthesizes hormones and hormone-releasing agents that affect distant organs and tissues.

26.75   The absorptive state is the condition of the body immediately after a meal, when the gastrointestinal tract is full.

26.77   Insulin.

26.79   Glucagon.

26.81   In the postabsorptive state, blood glucose is supplied by glycogenolysis. When starvation begins, glycogen is unavailable, and gluconeogenesis with the use of glucogenic amino acids from muscle provides the glucose.

26.83   It typically appears early in life and can be controlled by insulin replacement.

26.85   Gluconeogenesis with the use of glucogenic amino acids provides the glucose.

26.87   Because glucose is not available for energy production, fatty acid oxidation becomes the main source of ATP.

26.89   Nitrogen balance is achieved when the intake of protein nitrogen is equal to the loss of nitrogen in the urine and feces.

26.91   This vitamin is not ordinarily essential because it is synthesized in adequate amounts by intestinal bacterial flora. However, vitamin B_{12} is transported across the intestinal cell membrane as a complex with intrinsic factor; in persons who cannot synthesize this protein, the vitamin cannot enter the bloodstream through intestinal absorption.

26.93   The liver uses excess glucose for the synthesis of fatty acids and cholesterol.

26.95   If it were synthesized as the active proteolytic enzyme chymotrypsin, it would digest the pancreas itself.

26.97   Dietary lipid in the form of chylomicrons is hydrolyzed at the surface of capillary membranes within muscle and adipose tissue. The free fatty acids then penetrate the cells of the tissue to be stored as triacylglycerol or oxidized for energy.

26.99   The protein contained in the wheat is difficult to extract because it is located within an indigestible husk.

26.101 Sensory neurons spontaneously generate action potentials at frequencies that depend upon the intensity of an environmental stimulus.

APPENDIX

Solutions to *Organic and Biochemistry*

Chapters 1 and 2

Chapter 1. The Properties of Atoms and Molecules

1.1   Proton, mass 1.00728 amu, charge = +1

Neutron, mass 1.00867 amu, charge = 0

Electron, mass 0.0005486 amu, charge = −1

1.2   Atomic number = 16; element is sulfur.

1.3   Atomic mass includes the mass of neutrons.

1.4

Protons	Neutrons	Electrons	Mass (amu)	Element
19	20	**19**	39	**K**
34	**45**	**34**	79	**Se**
20	20	**20**	40	**Ca**
11	**12**	11	23	**Na**

1.5   It would be a cation with a charge = +3.

1.6   It would be an anion with a charge = -1.

1.7

Element	Group	Period
Li	I	2
Na	I	3

K	I	4
Rb	I	5
Cs	I	6

1.8

Element	Group	Period
Si	IV	3
Ge	IV	4
As	V	4
Sb	V	5

1.9   Li, Na, K, Rb, and Cs are metals because they are all on the left side of the Periodic Table.

1.10   Si, Ge, As, and Sb are elements that have properties of both metals and non-metals and are called metalloids.

1.11   Group I.

1.12   Group VII.

1.13   Group I. All elements possess 1 outer electron, ns^1.

1.14   Group VII. All elements possess 7 outer electrons, ns^2np^5.

1.15   Group II. All elements possess 2 outer electrons, ns^2.

1.16   Group VIII. All elements possess 8 outer electrons, ns^2np^6, except helium, $1s^2$.

1.17   True.

1.18   By losing, gaining, or sharing electrons.

1.19   The formation of either ionic or covalent bonds.

1.20   Group I forms +1 ions, Group VII forms −1 ions → For neutral compound ratio = 1:1

1.21   Group I forms +1 ions, Group VI forms −2 ions → For neutral compound ratio = 2:1

1.22   Group II forms +2 ions, Group VII form –1 ions → For neutral compound ratio = 1:2

1.23   Group III forms +3 ions, Group VI form –2 ions → For neutral compound ratio = 2:3

1.24   Calcium ion = Ca^{2+}. Any group VI element forms a –2 ion. Therefore to make a neutral compound, 1 calcium and 1 group VI element must be included in the compound. The first 3 elements in group VI are oxygen, O; sulfur, S; and selenium, Se. Thus,

    (a) CaO   (b) CaS   (c) CaSe

1.25   Aluminum ion = Al^{3+}. Any group VI element forms a –2 ion. Therefore to make a neutral compound, two aluminums and three group VI elements must be included in the compound. The first three elements in group VI are oxygen, O; sulfur, S; and selenium, Se. Thus,

    (a) Al_2O_3 (b) Al_2S_3 (c) Al_2Se_3

1.26   Aluminum ion = Al^{3+}. Any group VII element forms a –1 ion. Therefore to make a neutral compound, one aluminum and three group VII elements must be included in the compound. The first four elements in group VII are fluorine, F; chlorine, Cl; bromine, Br; iodine, I. Thus,

    (a) AlF_3 (b) $AlCl_3$ (c) $AlBr_3$ (d) AlI_3

1.27   Procedure to draw Lewis dot structure:

(i) Place atoms and connect bonds.

(ii) Determine how many valence electrons are available from all atoms

(iii) Subtract two electrons for each bond drawn in part (i).

(iv) Distribute remaining electrons among the atoms. Result should be that every atom has eight electrons associated with it either from free pairs of electrons or covalent bonds.

CCl_4

(i)

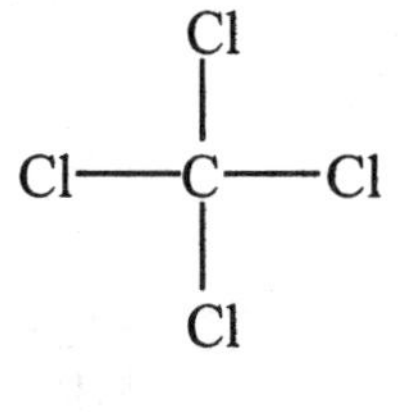

(ii) 4 electrons from the carbon + (7 electrons from each chlorine × 4) = 32 e⁻.

(iii) Covalent bonds use 8 electrons → 32 – 8 = 24 electrons remain.

(iv) Distribute 24 electrons among 4 Cl atoms (6 each) to give each Cl 8 electrons.

$$
\begin{array}{c}
\ddot{:}\ddot{\underset{\cdot\cdot}{Cl}}: \\
| \\
:\ddot{Cl}\!\!-\!\!C\!\!-\!\!\ddot{Cl}: \\
| \\
:\ddot{\underset{\cdot\cdot}{Cl}}:
\end{array}
$$

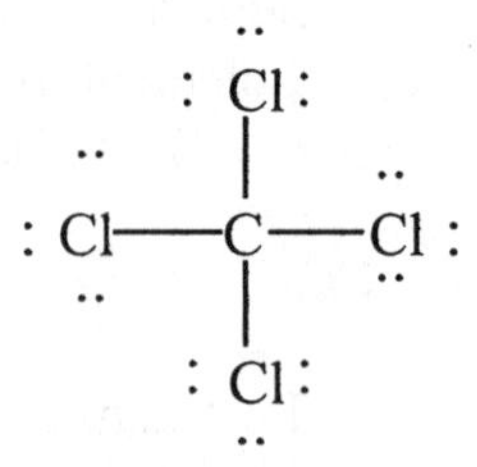

Check that all the atoms have 8 electrons associated with it. ---- ✓

1.28   Procedure to draw Lewis dot structure:

(i) Place atoms and connect bonds.

(ii) Determine how many valence electrons are available from all atoms

(iii) Subtract two electrons for each bond drawn in part (i).

(iv) Distribute remaining electrons among the atoms. Result should be that every atom has eight electrons associated with it either from free pairs of electrons or covalent bonds.

(i)

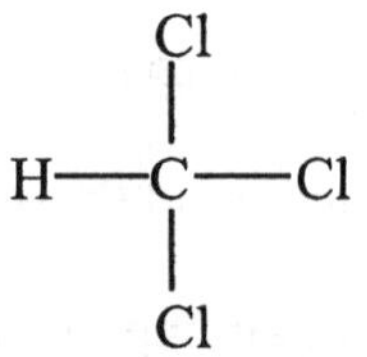

$$
\begin{array}{c}
Cl \\
| \\
H\!\!-\!\!C\!\!-\!\!Cl \\
| \\
Cl
\end{array}
$$

(ii) 4 electrons from the carbon + (7 electrons from each chlorine × 3) + 1 electron from the hydrogen = 26 e⁻.

(iii) Covalent bonds use 8 electrons → 26 – 8 = 18 electrons remain.

(iv) Distribute 18 electrons among 3 Cl atoms (6 each) to give each Cl 8 electrons.

$$
\begin{array}{c}
:\ddot{\underset{\cdot\cdot}{Cl}}: \\
| \\
H\!\!-\!\!C\!\!-\!\!\ddot{Cl}: \\
| \\
:\ddot{\underset{\cdot\cdot}{Cl}}:
\end{array}
$$

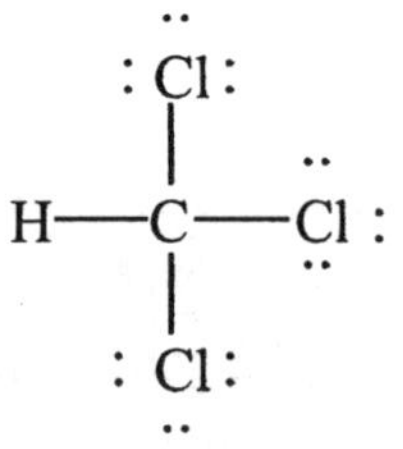

Check that each atom has eight electrons associated with it. ---- ✓

1.29   Using (i through iv) in 1.27:

(i)

$$H\!-\!S\!-\!H$$

(ii) 6 electrons from the sulfur + (1 electron from each hydrogen × 2) = 8 e⁻.

(iii) Covalent bonds use 4 electrons → 8 – 4 = 4 electrons remain.

(iv) Distribute four electrons among H and S. First put four electrons (two free pairs) on the sulfur to give it an octet.

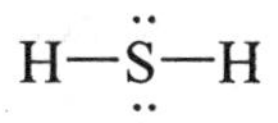

$$H\!-\!\overset{..}{\underset{..}{S}}\!-\!H$$

Check that each atom has eight electrons associated with it. ---- ✓

1.30   Using (i through iv) in 1.27:

(i)

$$Cl\!-\!\overset{\displaystyle |}{\underset{\displaystyle \underset{Cl}{|}}{N}}\!-\!Cl$$

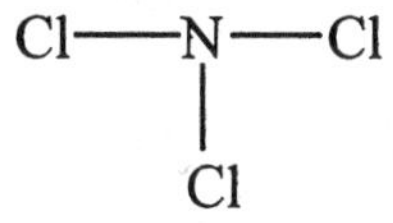

(ii) 5 electrons from the oxygen + (7 electrons from each chlorine × 3) = 26 e⁻.

(iii) Covalent bonds use 6 electrons → 26 – 6 = 20 electrons remain.

(iv) Distribute 20 electrons among Cl and N. First put two electrons (one free pair) on the nitrogen to give it an octet. That leaves 18 electrons, distribute these electrons among three Cl atom (six each) to give each Cl eight electrons.

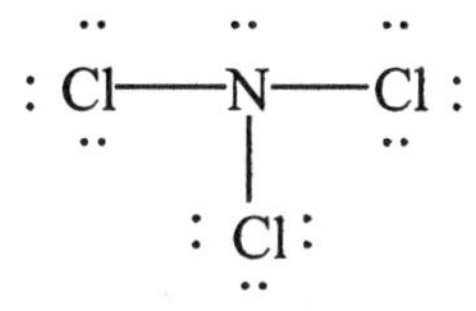

$$:\!\overset{..}{Cl}\!-\!\overset{..}{\underset{\displaystyle \underset{:\overset{..}{Cl}:}{|}}{N}}\!-\!\overset{..}{Cl}\!:$$

Check that each atom has eight electrons associated with it. ---- ✓

1.31   Using (i through iv) in 1.27:

(i)

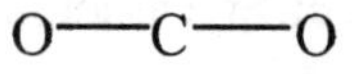

$$O\!-\!C\!-\!O$$

(ii) 4 electrons from the carbon + (6 electrons from each oxygen × 2) = 16 e⁻.

(iii) Covalent bonds use 4 electrons → 16 – 4 = 12 electrons remain.

(iv) Distribute 12 electrons among C and O. First put four electrons (two free pairs) on the carbon to give it an octet. That leaves eight electrons, distribute these electrons among two O atoms (four each) to give:

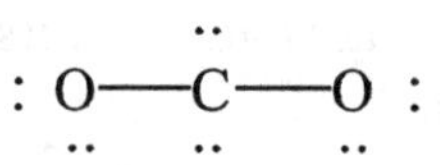

$$: O\!-\!C\!-\!O :$$

Check that each atom has eight electrons associated with it. ---- **NO**

Both oxygens have only six electrons. To remedy this, we need to incorporate multiple (double bonds). When there are not enough electrons to fill all atoms to an octet, changing single bonds to double bonds will result in more sharing of electrons.

Try again:

(i) Make bonds between C and O double bonds:

$$O\!=\!C\!=\!O$$

(ii & iii) Covalent bonds use 8 electrons → 16 – 8 = 8 electrons remain.

(iv) Distribute remaining electrons between the two oxygens as the carbon has eight electrons associated with it.

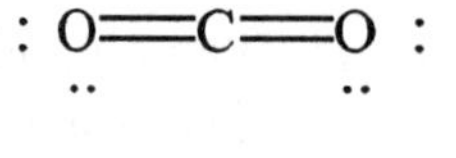

$$: O\!=\!C\!=\!O :$$

Check that each atom has eight electrons associated with it. ---- ✓

1.32   Using (i through iv) in 1.27:

(i)

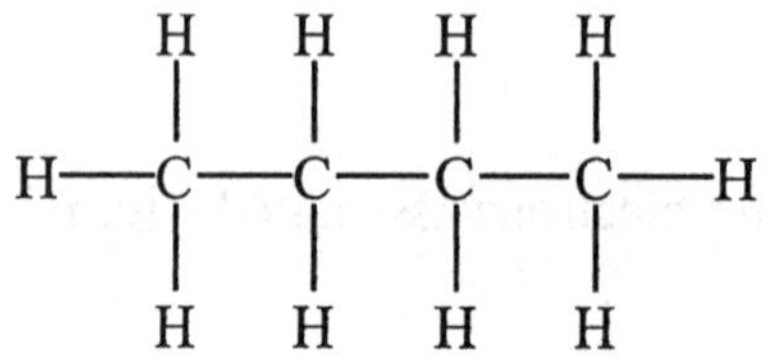

(ii) (4 electrons from the carbon × 4) + (1 electron from each hydrogen × 10) = 26 e⁻.

(iii) Covalent bonds use 26 electrons → 26 – 26 = 0 electrons remain.

(iv) No more free electrons, this structure is the Lewis dot structure:

$$
\begin{array}{ccccccc}
& H & & H & & H & & H \\
& | & & | & & | & & | \\
H- & C & - & C & - & C & - & C & -H \\
& | & & | & & | & & | \\
& H & & H & & H & & H
\end{array}
$$

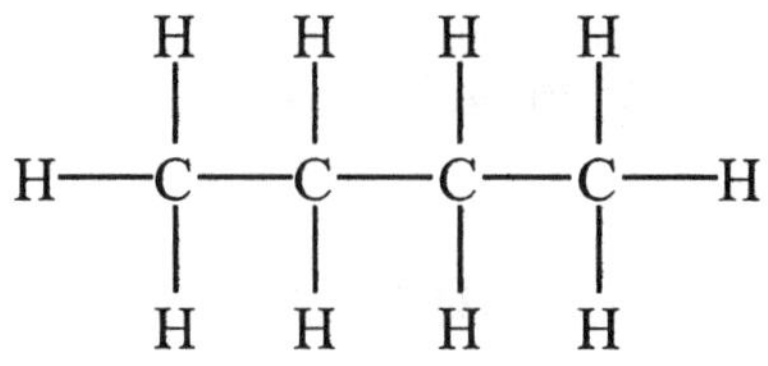

Check that each atom has eight electrons associated with it. ---- ✓

1.33   VSEPR provides information on the three-dimensional structure of a molecule by taking into account the fact that electrons have a negative charge. Therefore each pair of electrons in an atom wants to be as far away from all other pairs of electrons. In this theory "pairs of electrons" means either non-bonded pairs of electrons or covalent bonds. Furthermore, single, double, and triple bonds all count as a single "pair of electrons."

Therefore, before using VSEPR to determine the three-dimensional arrangement of the bonds/non-bonded electron pairs, we must first determine the Lewis dot structures. We will use the procedure outlined in Ex. 1.27.

(i)

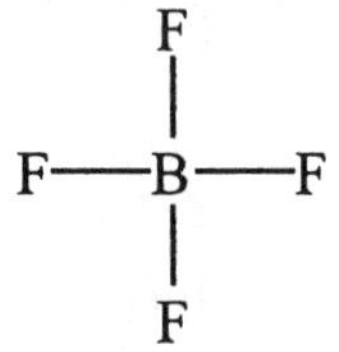

$$
\begin{array}{ccc}
& F & \\
& | & \\
F- & B & -F \\
& | & \\
& F &
\end{array}
$$

(ii) 3 electrons from the Boron + (7 electrons from each fluorine × 4) + 1 extra electron from the (–) charge = 32 e⁻.

(iii) Covalent bonds use 8 electrons → 32 – 8 = 24 electrons remain.

(iv) Distribute 24 electrons among four F (six each) to give:

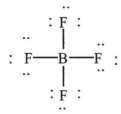

Therefore, the central atom is the B and it has four bonds, which equals four "pairs of electrons" around it. These four bonds must arrange so that they are as far away from each other as possible. This is accomplished by having the four fluorine atoms situated at the corners of a tetrahedron, with the boron atom at the center.

1.34   From (1.29), the Lewis dot structure of SH_2 is

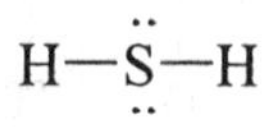

$$H-\overset{\cdot\cdot}{\underset{\cdot\cdot}{S}}-H$$

Therefore, the central atom is the S and it has two bonds and two non-bonded pairs, which equals four "pairs of electrons" around it. These four "pairs" must arrange so that they are as far away from each other as possible. This is accomplished by having the two hydrogen atoms and the two non-bonded pairs situated at the corners of a tetrahedron. Thus, the H–S–H bonds will form a bent molecule.

1.35   The Lewis dot structures are taken from Exercises 1.27, 1.33, and 1.31 for parts a, b, and c respectively.

(a)

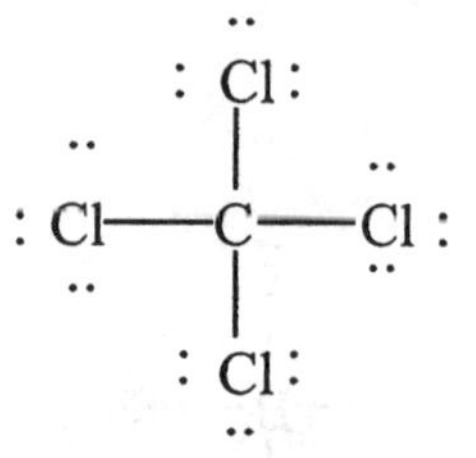

$$
\begin{array}{ccc}
 & :\overset{\cdot\cdot}{\underset{}{Cl}}: & \\
 & | & \\
:Cl\!\!-\!\!&C&\!\!-\!\!Cl: \\
 & | & \\
 & :\underset{\cdot\cdot}{Cl}: & \\
\end{array}
$$

From this structure and VSEPR, CCl_4 forms a tetrahedron with each C–Cl bond going from the center to the corner. Even though each C–Cl bond is polar, this arrangement results in a cancellation of these polar contributions to give a molecule with no dipole moment.

(b)

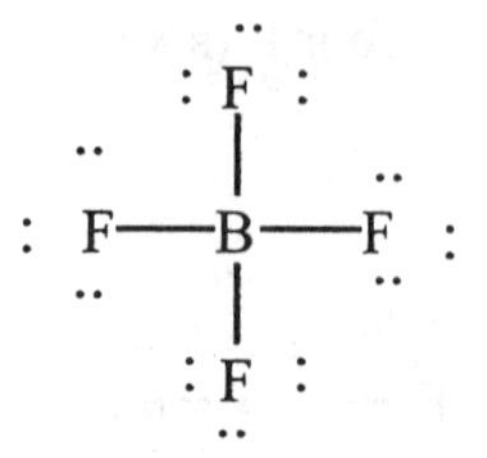

$$
\begin{array}{ccc}
 & :\overset{\cdot\cdot}{\underset{}{F}}: & \\
 & | & \\
:F\!\!-\!\!&B&\!\!-\!\!F: \\
 & | & \\
 & :\underset{\cdot\cdot}{F}: & \\
\end{array}
$$

From this structure and VSEPR, BF₄ forms a tetrahedron with each B–F bond going from the center to the corner. Even though each B–F bond is polar, this arrangement results in a cancellation of these polar contributions to give a molecule with no dipole moment.

(c)

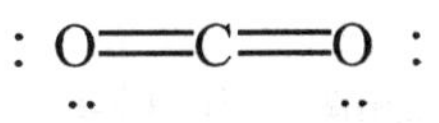

$$: O\!\!=\!\!\!=\!\!C\!\!=\!\!\!=\!\!O :$$

From this structure and VSEPR, CO_2 forms a linear molecule with each C=O 180 ° from each other. Even though each C–O– bond is polar, this arrangement results in a cancellation of these polar contributions to give a molecule with no dipole moment.

1.36   The Lewis dot structures are taken from Exercises 1.28, 1.30, and example 1.29 for parts a, b, and c respectively.

(a)

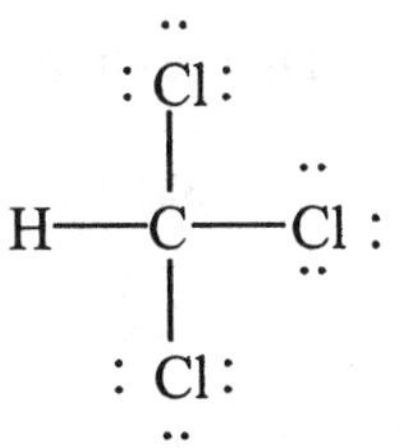

From this structure and VSEPR, $CHCl_3$ forms a tetrahedron with each C–Cl (or CH) bond going from the center to the corner. Because each C–Cl and CH bond is polar and of different strengths, this arrangement will not cancel out the polar contribution of each bond and therefore, results in a molecule with a dipole moment.

(b)

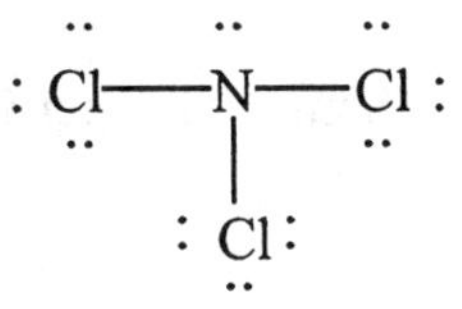

From this structure and VSEPR, NCl_3 forms a trigonal pyramid with each N–Cl bond going from the apex to the bottom. Because each N–Cl bond is polar, this arrangement results in a molecule with a dipole moment.

(c)

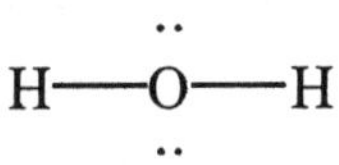

From this structure and VSEPR, H_2O forms a bent molecule. Because each H–O bond is polar, this arrangement results in a molecule with a dipole moment.

1.37   The four C–Cl polar bonds are arranged symmetrically (tetrahedrally) with carbon at the center of the tetrahedron. The symmetric arrangement of dipoles results in a net dipole moment of zero.

1.38   Carbon dioxide is a linear molecule. The C–O dipoles are directly opposed to each other, so the net dipole moment is zero.

1.39

	London force	Dipole-dipole	Hydrogen bond
CH_4	Yes	No	No
$CHCl_3$	Yes	Yes	No
NH_3	Yes	Yes	Yes

1.40

	London force	Dipole-dipole	Hydrogen bond
CCl_4	Yes	No	No
CHF_3	Yes	Yes	No
H_2O	Yes	Yes	Yes

1.41   Acetic acid molecules form hydrogen bonds among themselves and also to water. The attractive forces between acetic acid and water are similar, and therefore acetic acid will dissolve in water.

1.42   Pentane molecules can interact only by London forces while water can form hydrogen bonds. The attractive forces between pentane molecules and water are very different and will therefore not form solutions.

1.43   They have virtually identical chemical properties due to similar outer shell electronic structure.

1.44   Sulfur is a nonmetal.

1.45   Calcium is a metal.

1.46   They are inert and do not readily react to form compounds because they have a full outer electron shell.

1.47   In chemical reactions, atoms tend to attain the noble gas outer shell configuration of eight electrons.

1.48   The statement is incorrect; there are no more than eight elements that have equal numbers of protons, electrons, and neutrons: He, C, N, O, Ne, Si, S, and Ca.

1.49   The chemical properties are identical, but the nucleus of carbon-12 contains six neutrons and that of carbon-13 contains seven neutrons.

1.50   Total percent = 100; isotope at 26 amu = X%; isotope at 25 amu = (100 – X)%

$$100\left[\frac{X}{100}(26) + \frac{100-X}{100}(25) = 25.6\right]$$

26X – 25X + 2500 = 2650

X = 60, 100 – X = 40

Isotope at 26 amu = 60%

Isotope at 25 amu = 40%

1.51   Using (i through iv) in 1.27:

(i)

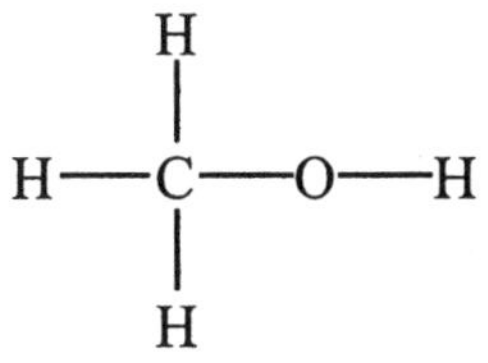

(ii) 4 electrons from the carbon + (1 electron from each hydrogen × 4) + 6 electrons from the oxygen = 14 e⁻.

(iii) Covalent bonds use 10 electrons → 14 – 10 = 4 electrons remain.

(iv) Inspection shows that the carbon and hydrogens have their outer shell full while oxygen is still short. Therefore distribute four electrons to the O atom to:

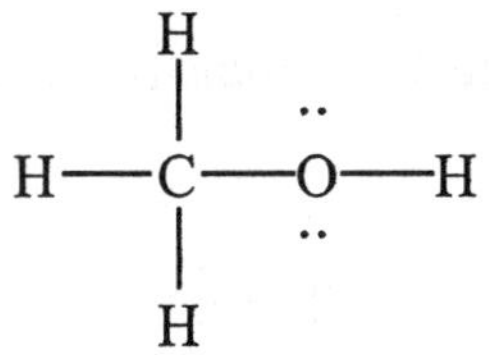

Check that each atom has eight electrons associated with it. ----   ✓

1.52   London dispersion forces are the result of temporary dipoles. A larger number of electrons in a molecule will result in a larger temporary dipole. Therefore, the greater the molecular mass, the stronger the London forces between molecules.

1.53   Elements that appear in the same group in the periodic table will exhibit similar chemical properties.

1.54   (a) From Table 2.8, hydrogen has one electron in the $1s$ subshell.

(b) From Table 2.8, neon has its outer shell completely full.

(c) From Table 2.8, sodium's outer subshell is the $3s$, and this atom has one atom in this s subshell, which makes it half full.

(d) From Table 2.8, nitrogen's outer subshell is the $2p$, which is full.

1.55   Electronegativity increases as you go across and up the periodic table, thus

(a) C < N < O

(b) Sr < Ca < Mg

(c) Pb < Ge < C

(d) Po < Te < Se

1.56   Using (i through iv) in 1.27:

(a) $PbCl_3$

(i)

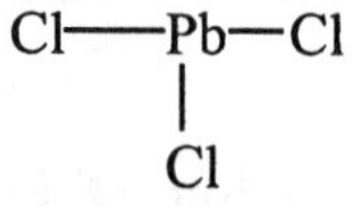

$$Cl—Pb—Cl$$
$$|$$
$$Cl$$

(ii) 4 electrons from the Pb + (7 electrons from each chlorine × 4) = 32 e⁻.

(iii) Covalent bonds use 6 electrons → 32 – 6 = 26 electrons remain.

(iv) Distribute 24 electrons among 4 Cl atoms (6 each) to give each Cl 8 electrons and 2 electrons to Pb.

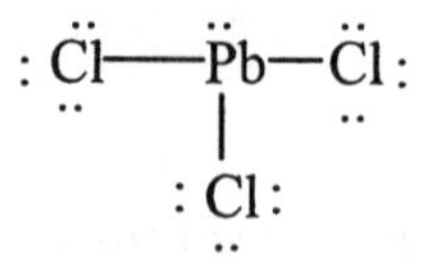

$$: \ddot{Cl}—\ddot{Pb}—\ddot{Cl} :$$
$$|$$
$$: \ddot{Cl} :$$

Check that all the atoms have 8 electrons associated with it.   ----   ✓

(b) $SO_4{}^{2-}$

(i)

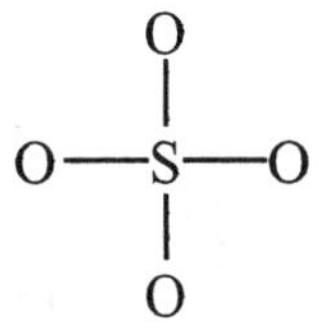

$$O-S-O$$

with O above and below the S (structure):

O
|
O—S—O
|
O

(ii) 6 electrons from the S + (6 electrons from each oxygen × 4) + 2 electrons for the 2–charge = 32 e⁻.

(iii) Covalent bonds use 8 electrons → 32 – 8 = 24 electrons remain.

(iv) Distribute 24 electrons among 4 O atoms (6 each) to give each O 8 electrons

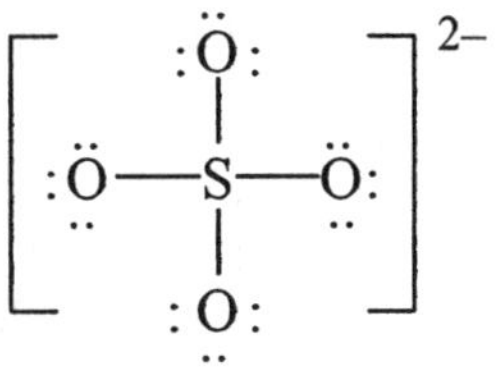

$$\left[\begin{array}{c} :\ddot{O}: \\ | \\ :\ddot{O}-S-\ddot{O}: \\ | \\ :\ddot{O}: \end{array} \right]^{2-}$$

Check that all the atoms have 8 electrons associated with it. ---- ✓

(c) PO_4^{3-}

(i)

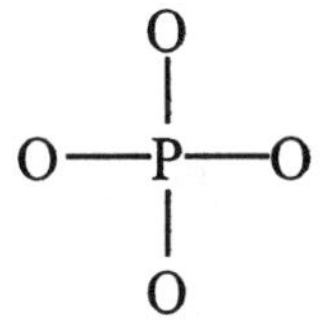

O
|
O—P—O
|
O

(ii) 5 electrons from the P + (6 electrons from each oxygen × 4) + 3 electrons for the 3–charge = 32 e⁻.

(iii) Covalent bonds use 8 electrons → 32 – 8 = 24 electrons remain.

(iv) Distribute 24 electrons among 4 O atoms (6 each) to give each O 8 electrons

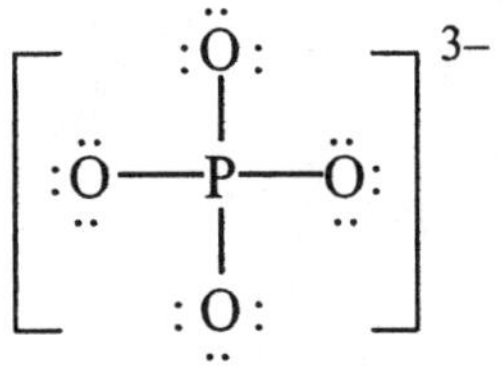

$$\left[\begin{array}{c} :\ddot{O}: \\ | \\ :\ddot{O}-P-\ddot{O}: \\ | \\ :\ddot{O}: \end{array} \right]^{3-}$$

Check that all the atoms have 8 electrons associated with it. ---- ✓

(d) ClO_4^{-}

(i)

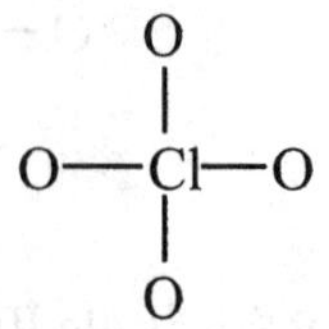

$$
\begin{array}{c}
\text{O} \\
| \\
\text{O—Cl—O} \\
| \\
\text{O}
\end{array}
$$

(ii) 7 electrons from the Cl + (6 electrons from each oxygen × 4) + 1 electrons for the – charge = 32 e⁻.

(iii) Covalent bonds use 8 electrons → 32 – 8 = 24 electrons remain.

(iv) Distribute 24 electrons among 4 O atoms (6 each) to give each O 8 electrons

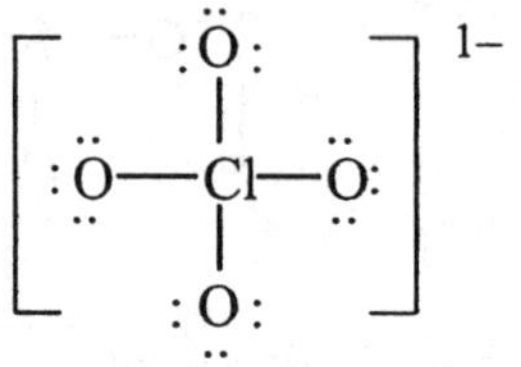

$$
\left[
\begin{array}{c}
:\ddot{\text{O}}: \\
| \\
:\ddot{\text{O}}—\text{Cl}—\ddot{\text{O}}: \\
| \\
:\ddot{\text{O}}:
\end{array}
\right]^{1-}
$$

Check that all the atoms have 8 electrons associated with it. ---- ✓

Chapter 2. Chemical Change

2.1 From Table 2.1:

(a) AgCl is not soluble in H_2O, therefore a precipitate forms:

$$Ag^+(aq) + Cl^-(aq) \rightarrow AgCl\ (s)$$

(b) $Mg(OH)_2$ is not soluble in H_2O, therefore a precipitate forms:

$$Mg^{2+}(aq) + 2\ OH^-(aq) \rightarrow Mg(OH)_2\ (s)$$

2.2 (a) $PbCl_2$ is not soluble in H_2O, therefore a precipitate forms:

$$Pb^{2+}(aq) + 2Cl^-(aq) \rightarrow PbCl_2\ (s)$$

(b) $Ca_3(PO_4)_2$ is not soluble in H_2O, therefore a precipitate forms:

$$3\ Ca^{2+}(aq) + PO_4^{3-}(aq) \rightarrow Ca_3(PO_4)_2\ (s)$$

2.3 (a) $BaSO_4$ is not soluble in H_2O, therefore a precipitate forms:

$$Ba^{2+}(aq) + SO_4^{2-}(aq) \rightarrow BaSO_4\ (s)$$

(b) $Al(OH)_3$ is not soluble in H_2O, therefore a precipitate forms:

$$Al^{3+}(aq) + OH^-(aq) \rightarrow Al(OH)_3\ (s)$$

2.4 (a) AgBr is not soluble in H_2O, therefore a precipitate forms:

$$Ag^+(aq) + Br^-(aq) \rightarrow AgBr\ (s)$$

(b) There is no reaction because all possible salts are soluble.

2.5 For reaction (a) to occur, the two reactants must have kinetic energies greater than 24 kJ while for reaction (b) to occur the two reactants must have kinetic energies greater than 53 kJ. At a given temperature, the percentage of molecules that have K.E. > 24 kJ will be greater than the percentage that have K.E. > 53 kJ. Therefore, the rate of reaction (a) will be greater than the rate of reaction (b).

2.6 The enzyme catalase is a catalyst that lowers the activation energy of the reaction, which in turn increases the reaction rate at a given temperature.

2.7 $\Delta H_{reaction} = E_{forward} - E_{back}$

$-21\ kJ = 37\ kJ - E_{back}$

$E_{back} = 58$ kJ

2.8 $\Delta H_{reaction} = E_{forward} - E_{back}$

12 kJ $= -46$ kJ $- E_{back}$

$E_{back} = -58$ kJ

2.9   No. The reaction takes place in an open system so that the CO_2 escapes to the atmosphere. Therefore, the CO_2 that is formed cannot react with CaO to allow the reverse reaction to occur. Therefore, the system can never come to equilibrium.

2.10   No. The reaction takes place in an open system so that the CO_2 escapes to the atmosphere. Therefore, the CO_2 that is formed cannot react with H_2O and $CaCl_2$ to allow the reverse reaction to occur. Therefore, the system can never come to equilibrium.

2.11   Forward reaction: $CaCO_3$ (s) $\rightarrow$ CaO (s) $+ CO_2$ (g)

Reverse reaction: CaO (s) $+ CO_2$ (g) $\rightarrow CaCO_3$ (s)

2.12   Forward reaction: CaO (s) $+ CO_2$ (g) $\rightarrow CaCO_3$ (s)

Reverse reaction: $CaCO_3$ (s) $\rightarrow$ CaO (s) $+ CO_2$ (g)

2.13 $K_{eq} = \dfrac{[HI]^2}{[H][I]} = \dfrac{(0.27)^2}{(0.86)(0.86)} = 0.099$

2.14 $K_{eq} = [NH_3][HCl] = [3.7 \times 10^{-3}][3.7 \times 10^{-3}] = 1.38 \times 10^{-5}$

2.15   Le Chatelier's principle states that when a reactant or product is removed from a reaction at equilibrium, the equilibrium shifts to replace the compound that was removed.

$$N_2 \text{ (g)} + 3H_2 \text{ (g)} \leftrightarrow 2NH_3$$

If NH_3 is removed, the reaction shifts to replace NH_3, which means that it shifts to the products.

2.16   Le Chatelier's principle states that when the temperature is raised in a reaction at equilibrium, the equilibrium shifts to lower the temperature. In this reaction, that means that the reaction shifts to the reactants.

2.17   Le Chatelier's principle states that when a reactant or product is added to a reaction at equilibrium, the equilibrium shifts to use up the compound that was added. In this reaction when Cl^- is added, the reaction shifts to use up more Cl^-; the reaction will shift to the products which will result in a color shift to blue.

2.18   Le Chatelier's principle states that when a reactant or product is added to a reaction at equilibrium, the equilibrium shifts to use up the compound that was added. In this reaction when Fe^{3+} is added, the reaction shifts to use up more Fe^{3+}; the reaction will shift to the products which will result in a color shift to red.

2.19 $COCl_2$ (g) $\leftrightarrow$ CO (g) + Cl_2 (g)

2.20 CH_4 (g) + H_2O (g) $\leftrightarrow$ CO (g) + 3 H_2 (g)

2.21 PCl_3 (g) + Cl_2 (g) $\leftrightarrow$ PCl_5 (g)

2.22   3 O_2 (g) $\leftrightarrow$ 2 O_3 (g)

2.23   The concentrations of pure liquids and solids do not appear in K_{eq}, therefore:

$$K_{eq} = \frac{[NO_2][NO_3]}{[N_2O_5]}$$

2.24   The concentrations of pure liquids and solids do not appear in K_{eq}, therefore:

$$K_{eq} = \frac{[N_2][H_2O]^2}{[NO]^2[H_2]^2}$$

2.25   The ion product, $K_w = [H_3O^+] \times [OH^-] = (1.00 \times 10^{-7}) \times (1.00 \times 10^{-7}) = 1.00 \times 10^{-14}$

2.26   In pure water $[H_3O^+] = [OH^-]$ and $K_w = [H_3O^+] \times [OH^-]$. Therefore, $K_w = [OH^-]^2 = [H_3O^+]^2 = 2.51 \times 10^{-14}$. $[H_3O^+] = [OH^-] = (K_w)^{1/2} = 1.58 \times 10^{-7}$.

2.27   A strong base is dissociated 100% in aqueous solution. Two examples are NaOH and LiOH.

2.28   A strong acid is dissociated 100% in aqueous solution. Two examples are HCl and H_2SO_4.

2.29   0.30 M HCl solution. HCl is a strong acid, which means that every HCl molecule dissociates into an H_3O^+ ion and a Cl^- ion. Thus, $[H_3O^+] = [Cl^-] = 0.30$ M.

2.30   0.28 M $HClO_3$ solution. $HClO_3$ is a strong acid, which means that every $HClO_3$ molecule dissociates into an H_3O^+ ion and a ClO^- ion. Thus, $[H_3O^+] = [ClO^-] = 0.28$ M.

2.31   The pH of pure water at 25°C is 7.00.

2.32   The pH of water at 37°C is 6.80. This is because the equilibrium reaction between two water molecules to form H_3O^+ and OH^- changes with temperature, and at 37°C, there is more H_3O^+ formed than at 25°C where the pH is 7.00.

2.33   pH = $-\log [H_3O^+] = -\log (0.010) = 2.00$.

2.34   pH = $-\log [H_3O^+] = -\log (0.0026) = 2.59$.

2.35 A weak acid is one that incompletely dissociates in water. Examples are given in Table 2.4, acetic acid and phosphoric acid.

2.36 A weak base is one that incompletely dissociates in water. Examples are given in Table 2.4, aniline and morphine.

2.37 (a) An acid is a compound that donates a proton, its conjugate base is the compound that is formed upon loss of a hydrogen. Both HNO_2 and H_3O^+ give up a proton to form NO_2^- and H_2O respectively. Therefore HNO_2 and NO_2^- are one conjugate acid-base pair, while H_3O^+ and H_2O are another.

(b) An acid is a compound that donates a proton, its conjugate base is the compound that is formed upon loss of a hydrogen. Both $H_2PO_4^-$ and H_3O^+ give up a proton to form HPO_4^{2-} and H_2O respectively. Therefore $H_2PO_4^-$ and HPO_4^{2-} are one conjugate acid-base pair, while H_3O^+ and H_2O are another.

2.38 (a) An acid is a compound that donates a proton, its conjugate base is the compound that is formed upon loss of a hydrogen. Both $HCOOH$ and NH_4^+ give up a proton to form $HCOO^-$ and NH_3 respectively. Therefore $HCOOH$ and $HCOO^-$ are one conjugate acid-base pair, while NH_4^+ and NH_3 are another.

(b) An acid is a compound that donates a proton, its conjugate base is the compound that is formed upon loss of a hydrogen. Both H_2O and H_3O^+ give up a proton to form OH^- and H_2O respectively. Therefore H_2O and OH^- are one conjugate acid-base pair, while H_3O^+ and H_2O are another.

2.39 Because the presence of the conjugate acid allows the solution to neutralize the addition of a base while the presence of the conjugate base allows the solution to neutralize the addition of a base. Because it is a conjugate acid-base pair, the solution remains an a constant pH.

2.40 $$ pH = pK_a + \log\left(\frac{proton\ acceptor}{proton\ donor}\right) = \log\left(\frac{0.0080}{0.0060}\right) = 4.76 + 0.13 = 4.89 $$

2.41 $$ pH = pK_a + \log\left(\frac{proton\ acceptor}{proton\ donor}\right) = \log\left(\frac{0.070}{0.070}\right) = 3.75 + 0 = 3.75 $$

2.42 $$ pH = pK_a + \log\left(\frac{proton\ acceptor}{proton\ donor}\right) = \log\left(\frac{0.075}{0.050}\right) = 7.20 + 0.18 = 7.38 $$

2.43 $$ pH = pK_a + \log\left(\frac{proton\ acceptor}{proton\ donor}\right) = \log\left(\frac{0.050}{0.075}\right) = 7.20 - 0.18 = 7.02 $$

2.44 (a) 0.0031 M HNO_3 solution. HNO_3 is a strong acid, which means that every HNO_3 molecule dissociates into an H_3O^+ ion and a NO_3^- ion. Thus, $[H_3O^+] = 0.0031$ M.

$$pH = -\log [H_3O^+] = -\log (0.0031) = 2.50$$

(b) 1.0 M HCl solution. HCl is a strong acid, which means that every HCl molecule dissociates into an H_3O^+ ion and a Cl^- ion. Thus, $[H_3O^+] = 1.0$ M.

$$pH = -\log [H_3O^+] = -\log (1.0) = 0.00$$

2.45   (a) 0.0069 M HI solution. HI is a strong acid, which means that every HI molecule dissociates into an H_3O^+ ion and an I^- ion. Thus, $[H_3O^+] = 0.0069$ M.

$$pH = -\log [H_3O^+] = -\log (0.0069) = 2.16$$

(b) 0.019 M HBr solution. HNO_3 is a strong acid, which means that every HBr molecule dissociates into an H_3O^+ ion and a Br^- ion. Thus, $[H_3O^+] = 0.0031$ M.

$$pH = -\log [H_3O^+] = -\log (0.019) = 1.72$$

2.46   The combination of a weak acid and the salt of its conjugate base is a buffer solution. The pH of a buffer is

$$pH = pK_a + \log\left(\frac{proton\ acceptor}{proton\ donor}\right) = \log\left(\frac{0.075}{0.050}\right) = 6.35 + .13 = 6.48$$

2.47   The combination of a weak acid and the salt of its conjugate base is a buffer solution. The pH of a buffer is

$$pH = pK_a + \log\left(\frac{proton\ acceptor}{proton\ donor}\right) = 8.2 = 7.2 + x$$

$$x = 1.0 = \log\left(\frac{proton\ acceptor}{proton\ donor}\right) \rightarrow HPO_4^{2-}/H_2PO_4^- = 10/1$$

2.48 $pH = -\log [H_3O^+] = 3.20 \rightarrow [H_3O^+] = 6.31 \times 10^{-4}$

HCl is a strong acid, which means that every HCl molecule dissociates into an H_3O^+ ion and a Cl^- ion. Thus, $[H_3O^+] = [HCl] = 6.31 \times 10^{-4}$ M.

2.49   (a) HCl is a strong acid, which means that every HCl molecule dissociates into an H_3O^+ ion and a Cl^- ion. Thus, $[H_3O^+] = [HCl] = .0034$ M. It is also known that

$$[H_3O^+] \times [OH^-] = 1.00 \times 10^{-14} \rightarrow \left[OH^-\right] = \frac{1.00 \times 10^{-14}}{0.0034} = 3.0 \times 10^{-12}$$

$$pH = -\log[H_3O^+] = 2.47$$

(b) HNO_3 is a strong acid, which means that every HNO_3 molecule dissociates into an H_3O^+ ion and a NO_3^- ion. Thus, $[H_3O^+] = [HNO_3] = 0.025$ M. It is also known that

$$[H_3O^+] \times [OH^-] = 1.00 \times 10^{-14} \rightarrow \left[OH^-\right] = \frac{1.00 \times 10^{-14}}{0.025} = 4.0 \times 10^{-13}$$

$$pH = -\log[H_3O^+] = 1.60$$

2.50   A combination of a weak acid and the salt of its conjugate base is a buffer solution. Thus examples include acetic acid and potassium acetate, formic acid and sodium formate, and carbonic acid and sodium bicarbonate.

2.51   A catalyst is a chemical that increases the rate of a reaction, but is not used up in the overall reaction. Therefore, NO is the catalyst.

2.52 $Mg(OH)_2$ (s) $\leftrightarrow$ Mg^{2+} (aq) + 2 OH^- (aq)

If H^+ is added to this reaction, it will react with OH^- as

$$H^+ + OH^- \leftrightarrow H_2O$$

The net result is that OH^- is lost from the first equilibrium. Thus, that equilibrium will respond by trying to replace that OH^- by pushing that reaction to the right.

2.53   If a system in an equilibrium state is disturbed, the system will adjust to neutralize that disturbance and restore the system to equilibrium.

2.54   The ionization of water is an equilibrium reaction:

$$H_2O \leftrightarrow H_3O^+ + OH^-$$

The fact that the ion product of water increases with temperature means that this equilibrium is shifted to the right (products) with an increase in temperature. According to Le Chatelier's principle, when an equilibrium is perturbed, it will shift to bring the equilibrium back to its initial conditions. Therefore, when the temperature is increased, this reaction will shift to the right, which lowers the temperature. A reaction that lowers the temperature as it goes toward its products is endothermic.

2.55   Ethanol is in excess; therefore it is considered to be the solvent, and water is the solute.

2.56 $V_1M_1 = V_2M_2$

$V_1 = ?$ $M_1 = 0.300$ M

$V_2 = 135$ mL $M_2 = 0.180$ M

$$V_1 = \frac{135 \text{ mL} \times 0.180 \text{ M}}{0.300 \text{ M}} = 81.0 \text{ mL of original solution needed.}$$

Therefore, she needs 81.0 mL of original solution to make 135 mL of the final solution. Since she only has 70.0 mL, she will not have enough.

2.57 $13.9 \text{ g} \times \dfrac{1 \text{ mol}}{95.3 \text{ g}} = 0.146 \text{ mol}; \ 0.146 \text{ mol}/0.225 \text{ L} = 0.649 \text{ M}.$

2.58   The square brackets around chemical components in an equilibrium constant expression denote their molar concentrations at equilibrium.

2.59   A conjugate acid-base pair is a weak acid and the basic anion that results from its dissociation.

2.60   Both water and hydrogen sulfide can form dipole-dipole and London secondary attractive forces, but water can form hydrogen bonds, hydrogen sulfide cannot. The strong hydrogen bonds lead to water being a solid at the same temperature that hydrogen sulfide exists as a gas.

2.61   When a substance melts, the intermolecular forces are loosened so that the order is reduced, but the molecules are still largely in contact with one another, similar to a solid. Additionally, the liquid is about as incompressible as a solid.

2.62   (a) LeChatelier's Principle states that when a reactant or product is removed from a reaction at equilibrium, the equilibrium shifts to replace the compound that was removed, thus this reaction will shift to create more carbon dioxide and thus to the right (products).

(b) LeChatelier's Principle states that when a reactant or product is added to a reaction at equilibrium, the equilibrium shifts to use up the compound that was added, thus this reaction will shift to use up the extra water vapor and thus to the right (products).

(c) LeChatelier's Principle states that when the pressure is increased in a reaction at equilibrium, the reaction will shift to decrease the pressure. However, in this reaction the number of moles of gas is equal on both sides of the reaction, and thus this increase in pressure does not change the equilibrium.

2.63   An overall equilibrium constant for a sequence of reactions can be calculated only if the products of the first reaction are the reactant of the next reaction, and so on, for each succeeding reaction.

2.64   When 50% of the acid has been neutralized, the concentration of acid (e.g., acetic acid) in solution is equal to the concentration of the acid anion (e.g., acetate).

According to the Henderson-Hasselbalch equation, the logarithm is then equal to zero, and the pH is equal to the pK_a.

2.65   The pH at the equivalence point of the titration in Exercise 2.64 could not be 7.00. At the equivalence point, the acid has been completely neutralized and the resulting solution is that of its salt. Because the salt consists of a basic anion, hydrolysis of the anion will cause the solution to be basic. The solution's pH will be greater than 7.00.

2.66 $pH = -\log[H_3O^+]$, therefore, the larger the pH, the smaller the concentration of hydronium ion. As the concentration of hydronium ion increases with dissociation constant, the compound with the smallest dissociation constant will have the highest pH and the dissociation constants will increase with decreasing pH.

$$6.65 < 4.82 < 3.41 < 2.85$$

2.67 Catalysts provide an alternate path for a reaction of lower E_a, which results in faster rate. ΔH is unchanged because the reactants and products are the same.

2.68 Catalysts provide an alternate path for a reaction of lower E_a, which results in faster rate. ΔH is unchanged because the reactants and products are the same.